DISTRIBUTION THEORY AND APPLICATIONS

SERIES ON CONCRETE AND APPLICABLE MATHEMATICS

ISSN: 1793-1142

Series Editor: Professor George A. Anastassiou
Department of Mathematical Sciences
The University of Memphis
Memphis, TN 38152, USA

Series on Concrete and Applicable Mathematics – Vol. 9

DISTRIBUTION THEORY AND APPLICATIONS

Abdellah El Kinani

Mohamed Oudadess

Ecole Normale Supérieure de Takaddoum, Morocco

 World Scientific

NEW JERSEY · LONDON · SINGAPORE · BEIJING · SHANGHAI · HONG KONG · TAIPEI · CHENNAI

Published by

World Scientific Publishing Co. Pte. Ltd.

5 Toh Tuck Link, Singapore 596224

USA office: 27 Warren Street, Suite 401-402, Hackensack, NJ 07601

UK office: 57 Shelton Street, Covent Garden, London WC2H 9HE

British Library Cataloguing-in-Publication Data
A catalogue record for this book is available from the British Library.

Series on Concrete and Applicable Mathematics — Vol. 9
DISTRIBUTION THEORY AND APPLICATIONS

ISBN-13 978-981-4304-91-7
ISBN-10 981-4304-91-3

Printed in Singapore.

to those, among our teachers, who have been sincere, from the primary school to the university; each one, according to what he has given to us

Preface

During the golden era of Functional Analysis around the sixties of the last century, the Theory of Distributions was already a *legend*, being considered as the paradise of the theory of locally convex (topological vector) spaces, the latter brought to that perfection by the famous Bourbaki group, in France, among them Laurent Schwartz himself and Jean Dieudonné, yet Alexander Grothendieck, George Mackey in the United States, Gottfried Köthe in Germany, just to mention quite a few of the main contributors to nowadays theory of locally convex spaces.

It is certain that we have already a plethora of books on the subject matter of the work in hand, that is the *Theory of* (the so-called Schwartz) *Distributions*. Of course the fact is not surprising when taking into account the importance of the subject and its vast amount of applications throughout mathematics, included here (mathematical) physics, a subject made such(!) after its formal (rigorous) shaping in the hands of great masters of the matter, as e.g. the late L. Schwartz, J. Dieudonné, J.-J. Lions, to name but a few.

However, the present account offered by the authors has its own merit, this mainly due to the clarity of exposition and the elementary way of presenting it, yet in a quite rigorous manner, let alone its thoroughness. These striking issues of the present writing are just the outcome of the experience, from the part of the authors, of many years teaching the subject in the class accompanied by the relevant tests and reactions from the audience. The authors, well-known experts in Topological Algebra Theory, a flourishing subject of present-day Functional Analysis, hence connoisseurs as well, of ramified theoretical details of the theory, as the latter became in the hands of the masters of the subject, succeed in presenting their theme precisely, and also in a quite comprehensible and down to earth manner. This makes the treatise accessible even to undergraduates as it should actually be the case, given its significance in mathematics, as mentioned in the beginning, and also in physics, theoretical and experimental alike.

Nowadays, we have already generalized aspects of the above classical theory of distributions, aiming at dealing with the awkward for the applications problem of *"multiplication of distributions"* in that the (ordinary/point-wise) product in the (vector) space of distributions does not always fall in the same space, something already (re)marked by L. Schwartz himself and also by P. Levy. In that direction we have the work of F. Colombeau and E.E. Rosinger, as also of others, with applications again in physics, yet in problems of *quantum gravity*. Furthermore, the same theory of (Schwartz) distributions might be considered as the "paradise" of topological (non-normed (!)) algebras, for instance, the latter being also a thriving subject today, even in the standard operator-theory régime, the classical acropolis of Banach algebra theory. We thus meet here a vindication indeed of the prediction of the late M.A. Naimark, in that"... *the theory of generalized functions* [: distributions] *of Laurent Schwartz ... should play an essential rôle in* [the theory of] *topological* (... *non-normed) rings* [: algebras] *...",* the latter discipline being also viewed, as *important* [(!)] *for many applications* (the same author as before [the italics above is ours].

All told, this nice and really quite enjoyable work in hand offers to the reader a leisurely pedagogical, as well as, a quite rigorous introduction to the subject of generalized functions à la Schwartz, an important area of classical and of contemporary substance alike, in an elementary and clear-cut manner. But anyway, ... *what can be said at all, can be said clearly* (!) (the italics ours), to recall here L. Wittgenstein. I found the book in hand marvelously corresponding to the above maxim, so that I wish, and also can hope, of its being very well received in the mathematical/physical community today.

Athens
September, 2009

Anastasios Mallios

Contents

Introduction

Having taught several years in a row the course on distributions at the Ecole Normale Supérieure (Takaddoum) of Rabat (Morocco), we find it useful to write an introductory book in an accessible and self-contained context; presenting at the same time some useful strong tools of the theory. We have, in adequate situations, produced calculations on regular distributions in order to justify the introduction of certain notions. We have made an effort to make the proofs the more transparent possible. The theory of distributions has many applications and in various domains. Hence they do not, all of them, have a place in a first course. We have limited ourselves here to the utilization of Fourier and Laplace transformations in the resolution of some differential equations.

We indicate briefly the content of this work. Other comments are given at the beginning of the chapters.

The fundamental spaces, which play an essential role in the theory of distributions, are presented in Chapter 1. General properties of distributions are given in Chapter 2. We also study, in the latter, the first classical examples (regular distributions, measures, principal value of Cauchy, finite parts of Hadamard, ...). For more clarity, we have reserved a chapter to tensor products (Chapter 3). The convolution product is the subject of Chapter 4. The notion of an allowed family is also discussed. Diverse classical equations (partial differential equations with constant coefficients, finite difference equations with constant coefficients and Volterra equations) are seen as convolution equations in suitable algebras (said to be of convolution). For convenience and due to the importance of the Fourier transformation, we reserved Chapter 5 to it, and this in $L^1(\mathbb{R}^n)$. We also give, in the latter, the transfer theorem and the preparatory formula of Riesz. In Chapter 6, we present and study the space $\mathcal{S}(\mathbb{R}^n)$ of Schwartz, the elements of which are the rapidly decreasing functions of class C^∞ as well as their derivatives of any order. We also study the space of tempered distributions which is the topological dual $\mathcal{S}'(\mathbb{R}^n)$ of $\mathcal{S}(\mathbb{R}^n)$. At the end of this chapter the notion of ultradistribution is discussed. Structures of some distributions are described in Chapter 7. In the eighth chapter we determine Fourier ranges of some subspaces of $\mathcal{S}'(\mathbb{R}^n)$. Essential properties of Laplace transformation are discussed in Chapter 9. We also present simple applications of Laplace transform in the resolution of partial differential equations. Different notions of kernels (regular, regularizing and very regular) are the subject of Chapter 10. They are used in the resolution of partial differential equations. Finally, in Chapter 11, we introduce Sobolev spaces $H^s(\mathbb{R}^n)$, $s \in \mathbb{R}$, and we present the injection theorem of Sobolev-Rellich.

Discussions are provided at the end of each chapter. It is a kind of brief comments on the results in relation with others, as well as their importance. Indications are given concerning the presentation in comparison with books in the Bibliography. Ideas on the introduction of the notions, the choice of the methods and the driving of the proofs are pointed out whenever possible. We also suggest further readings from the references according to the content.

Rabat   Abdellah El Kinani
November, 2009   Mohamed Oudadess

Acknowledgements

Our chance is to have had Prof. A. Mallios as a friend. We have presented to him the French version of the manuscript. He took time to check it and then gave his approval. He suggested to publish it in French. We could not. He then asked to translate it into English. We have been reluctant, but he has friendly insisted. Furthermore he accepted to write a preface; we could not completely express our gratitude to Professor Mallios. Without his accompaniment during the whole process, this book would have never appeared, especially in English.

We feel greatly honored that Prof. J. Horváth has been interested in our work. He furthermore has generously put much time, twice, to carefully examine it. He has written down, in detail, so many remarks and suggestions which contributed to bring the manuscript to its present form. They concern both the content and the presentation. His contribution allowed a significant improvement of the initial version of the book.

We also offer our warm thanks to Prof. J. de Graaf for his kind willing to look at the manuscript. He has made several remarks and suggestions. He has detected, in particular, incorrect and/or misleading statements.

Our sincere thanks are also due to Prof. G. Anastassiou (Univ. of Memphis) for his lively interest in our work and his steady encouragement in the publication of the book. The same goes to Mrs Zhang Ji for the solicitude and celerity during the final stage of the editing.

We remember with gratitude the late Popi Bolioti for her so kind and generous care in finalizing the typing of the manuscript.

Our special thanks are due to our colleague Marina Haralampidou for her so kind and instrumental help in finalizing the typesetting. Yet our sincere thanks to Rosa Garderi for similar reasons. Our thanks are also offered to the Director of the Ecole Normale Supérieure, Professor Abdelatif Moquine, for his comprehension and support.

We also express our thanks to several classes of senior students who attended our course for their questions and their remarks which led us to clarify some aspects and/or avoid useless lengths and go directly to the purpose.

Chapter 1

Spaces of Functions

In this chapter we present spaces of functions which play an essential role in the theory of distributions. They are called fundamental spaces or test function spaces. They are endowed with suitable topologies said to be canonical ones. Finally, density of certain spaces in others are given.

1.1 The space $\mathcal{C}(\Omega)$

We begin with a space which plays an important role in many domains of analysis. For an open subset Ω of $\mathbb{R}^n$, put

$$\mathcal{C}(\Omega) = \{f : \Omega \longrightarrow \mathbb{C} \ / \ f \text{ is continuous}\},$$

where $\mathbb{C}$ is the field of complex numbers. It is a complex vector space. Endow it with the topology of uniform convergence on compact sets, defined by the family $(p_K)_K$ of seminorms, where K runs over the collection of compact subsets of Ω, given by

$$p_K(f) = \sup_{x \in K} |f(x)|.$$

Then $(\mathcal{C}(\Omega), (p_K)_K)$ is a separated (i.e. Hausdorff) locally convex space (*l.c.s.*, in brief). An important property of this space is that it is metrizable. Indeed, it is sufficient to notice that its topology can be defined by the sequence $(p_{K_l})_l$ of seminorms, where $(K_l)_l$ is an exhaustive sequence of compact subsets of Ω. We do have the following essential result.

Proposition 1.1. *The space* $(\mathcal{C}(\Omega), (p_K)_K)$ *is a complete metrizable l.c.s., so a Fréchet space.*

1

Proof.   Let $(f_p)_p$ be a Cauchy sequence in $\mathcal{C}(\Omega)$. Fix a compact subset K of Ω. One has

$$\forall \varepsilon > 0, \ \exists N_K : p, q \geq N_K \Longrightarrow \sup_{x \in K} |f_p(x) - f_q(x)| \leq \varepsilon,$$

so, for every $x \in K$, the sequence $(f_p(x))_p$ is Cauchy in $\mathbb{C}$ which is complete. Whence one gets a function f_K defined on K by $f_K(x) = \lim\limits_{p \to +\infty} f_p(x)$. In fact, one has uniform convergence on K and hence f_K is continuous on K. The family $(f_K)_K$ of functions allows to obtain a function f defined on Ω by $f_{|K} = f_K$. This function f is continuous. Finally f is, by its construction, the limit of the sequence $(f_p)_p$ for the topology of $\mathcal{C}(\Omega)$.

$\square$

1.2   The space $\mathcal{E}^m(\Omega)$

For $m \in \mathbb{N}^*$, put

$$\mathcal{C}^m(\Omega) = \{f : \Omega \longrightarrow \mathbb{C} \ / \ f \text{ is of class } \mathcal{C}^m\}.$$

It is a complex vector space. One endows it with a topology for which it is a Fréchet *l.c.s.*. The following notations are needed. If $j = (j_1, j_2, \ldots, j_n)$ is a multi-index, i.e., an n-tuple of positive integers, put

$$D^j = \left(\frac{\partial}{\partial x_1}\right)^{j_1} \cdots \left(\frac{\partial}{\partial x_n}\right)^{j_n} = \frac{\partial^{j_1}}{\partial x_1^{j_1}} \cdots \frac{\partial^{j_n}}{\partial x_n^{j_n}}$$

and call $|j| = j_1 + \cdots + j_n$ the length of j.

Consider on $\mathcal{C}^m(\Omega)$ the seminorms $p_{K,m}$, where K is any compact subset of Ω and

$$p_{K,m}(f) = \max_{|j| \leq m} \sup_{x \in K} \left|D^j f(x)\right|.$$

So it is endowed with a separated structure of a *l.c.s.*. We have the following.

Proposition 1.2. *The space* $(\mathcal{C}^m(\Omega), (p_{K,m})_K)$ *is a Fréchet space.*

Proof.   For the metrizability use the same argument as for $\mathcal{C}(\Omega)$. Concerning completeness, let $(f_p)_p$ be a Cauchy sequence in $\mathcal{C}^m(\Omega)$. For any $|j| \leq m$ and any compact subset K of Ω, one has

$$\sup_{x \in K} \left|D^j f_p(x) - D^j f_q(x)\right| \xrightarrow[p, q \to +\infty]{} 0.$$

So for any $|j| \leq m$, $(D^j f_p)_p$ is a Cauchy sequence in $\mathcal{C}(\Omega)$ which is complete. Hence $D^j f_p \underset{p \to +\infty}{\longrightarrow} g_j$ in $\mathcal{C}(\Omega)$. And in particular, for $|j| = 0$, one has $f_p \underset{p \to +\infty}{\longrightarrow} g_0 = f$ in $\mathcal{C}(\Omega)$. It remains now to see that $f \in \mathcal{C}^m(\Omega)$ and that $f_p \underset{p \to +\infty}{\longrightarrow} f$ in $\mathcal{C}^m(\Omega)$. For this, it is sufficient to show that $g_j = D^j f$, for any j with $|j| \leq m$. One argues by induction on $|j|$. For $|j| = 0$, we do have $g_0 = f$. Suppose the result is true for $|j| \leq r$ with $r \leq m$ and let $|j| = r+1$. Consider the smallest integer k such that $j_k \neq 0$. So

$$D^j f_p = \frac{\partial}{\partial x_k}(D^{j'} f_p) \quad \text{with} \quad |j'| = r.$$

To finish, apply successively the induction hypothesis and the following result (cf. [15]): If a sequence $(f_p)_p \subset \mathcal{C}^m(\Omega)$ converges to f in $\mathcal{C}(\Omega)$ and if the sequence $\left(\dfrac{\partial f_p}{\partial x_i}\right)_p$ converges to h_i in $\mathcal{C}(\Omega)$, then f is differentiable with respect to x_i and $h_i = \dfrac{\partial f}{\partial x_i}$. □

Remark 1.1. If we want the notation $\mathcal{C}^m(\Omega)$ to include the case $\mathcal{C}(\Omega)$, we should adopt the convention $\mathcal{C}^0(\Omega) = \mathcal{C}(\Omega)$. In the theory of distributions it is usual to write $\mathcal{E}^m(\Omega)$ instead of $\mathcal{C}^m(\Omega)$, with obviously $\mathcal{E}^0(\Omega) = \mathcal{C}(\Omega)$.

1.3 The space $\mathcal{E}(\Omega)$

Put

$$\mathcal{C}^\infty(\Omega) = \{f : \Omega \longrightarrow \mathbb{C} \ / \ f \text{ is of class } C^\infty\}.$$

It is a complex vector space. One endows it with a topology for which it is a Fréchet *l.c.s.*. In $\mathcal{C}^\infty(\Omega)$, consider the seminorms $p_{K,m}$, where m is any positive integer, K any compact subset of Ω, and

$$p_{K,m}(f) = \max_{|j| \leq m} \sup_{x \in K} |D^j f(x)|.$$

Depend on in comparison with the space $\mathcal{C}^m(\Omega)$, of the previous section, it is not only the compact set K that varies but also the positive integer m. The space $\mathcal{C}^\infty(\Omega)$ endowed with the family $(p_{K,m})_{K,m}$ of seminorms is a separated *l.c.s.* And we have the following:

Proposition 1.3. *The space* $(\mathcal{C}^\infty(\Omega), (p_{K,m})_{K,m})$ *is a Fréchet space.*

Proof. For the metrizability it is again the same argument as for $\mathcal{C}(\Omega)$. Concerning the completeness, let $(f_p)_p$ be a Cauchy sequence in $\mathcal{C}^\infty(\Omega)$. It is then so in each $\mathcal{C}^m(\Omega)$ in which it is convergent because these spaces are complete. So $f_p \underset{p\to+\infty}{\longrightarrow} h_m$ in $\mathcal{C}^m(\Omega)$, for every m. In fact h_m does not depend of m. Indeed if $m \neq m'$, one has in particular $f_p \underset{p\to+\infty}{\longrightarrow} h_m$ and $f_p \underset{p\to+\infty}{\longrightarrow} h_{m'}$ in $\mathcal{C}(\Omega)$. Therefore $h_m = h_{m'}$. If we put $f = h_0$, we then have $f_p \underset{p\to+\infty}{\longrightarrow} f$, in $\mathcal{C}^m(\Omega)$ for every m, i.e., $f_p \underset{p\to+\infty}{\longrightarrow} f$ in $\mathcal{C}^\infty(\Omega)$. $\square$

Remark 1.2. In the theory of distributions, it is customary to write $\mathcal{E}(\Omega)$ instead of $\mathcal{C}^\infty(\Omega)$.

1.4   The space $\mathcal{D}_K^m(\Omega)$

Recall that the support of a function $f : \Omega \longrightarrow \mathbb{C}$, denoted by *supp f*, is defined as follows

$$supp\, f = \overline{\{x \in \Omega : f(x) \neq 0\}}.$$

$\bar{A}$ denotes the closure of a set A. If K is a compact subset of Ω and m a positive integer, put

$$\mathcal{D}_K^m(\Omega) = \{f : \Omega \longrightarrow \mathbb{C} \;/\; f \in \mathcal{C}^m(\Omega) \text{ and } supp\, f \subset K\}.$$

It is a complex vector space. One endows it with the topology induced by that one of $\mathcal{E}^m(\Omega)$. It can be defined by the single seminorm $p_{K,m}$ given by

$$p_{K,m}(f) = \max_{|j|\leq m} \sup_{x\in K} \left|\mathcal{D}^j f(x)\right|.$$

In fact it is a norm. The same is sometimes denoted by $\|.\|_{K,m}$. One easily checks that $\mathcal{D}_K^m(\Omega)$ is closed in $\mathcal{E}^m(\Omega)$. So we have the following result.

Proposition 1.4. *The space* $\left(\mathcal{D}_K^m(\Omega), \|.\|_{K,m}\right)$ *is a Banach space.*

Remark 1.3. $\mathcal{D}_K^0(\Omega)$ is denoted by $\mathcal{K}_K(\Omega)$; so

$$\mathcal{K}_K(\Omega) = \{f : \Omega \longrightarrow \mathbb{C} \;/\; f \in \mathcal{C}(\Omega) \text{ and } supp\, f \subset K\}.$$

1.5   The space $\mathcal{D}_K(\Omega)$

Let K be a compact subset of $\mathbb{R}^n$ and put

$$\mathcal{D}_K(\Omega) = \{f : \Omega \longrightarrow \mathbb{C} \;/\; f \in \mathcal{C}^\infty(\Omega) \text{ and } supp\, f \subset K\}.$$

It is a complex vector space. One endows it with the topology induced on it by $\mathcal{E}(\Omega)$. Recall that the latter is defined by the family $(p_{K,m})_m$ of seminorms, m running over the set of positive integers, with

$$p_{K,m}(f) = \max_{|j| \leq m} \sup_{x \in K} \left| D^j f(x) \right|.$$

Each one is in fact a norm. One easily checks that $\mathcal{D}_K(\Omega)$ is closed in $\mathcal{E}(\Omega)$. Hence we have the following result.

Proposition 1.5. *The space* $(\mathcal{D}_K(\Omega), (p_{K,m})_m)$ *is a Fréchet space.*

Remark 1.4. If we want the notation $\mathcal{D}_K^m(\Omega)$ to include also the case $\mathcal{D}_K(\Omega)$, we are led to adopt the convention $\mathcal{D}_K^\infty(\Omega) = \mathcal{D}_K(\Omega)$. Henceforth, we will write $\mathcal{D}_K^m(\Omega)$, where $m \in \overline{\mathbb{N}} = \mathbb{N} \cup \{+\infty\}$.

1.6 The space $\mathcal{D}^m(\Omega)$

Given $m \in \overline{\mathbb{N}}$, put

$$\mathcal{D}^m(\Omega) = \{f : \Omega \longrightarrow \mathbb{C} \ / \ f \in C^m(\Omega) \text{ and } supp \, f \text{ is compact}\}.$$

We note that $\mathcal{D}^0(\Omega)$ is denoted by $\mathcal{K}(\Omega)$ and $\mathcal{D}^\infty(\Omega)$ by $\mathcal{D}(\Omega)$.

Here are some examples.

Example 1.1. For any $c > 0$, let $f : \mathbb{R} \longrightarrow \mathbb{R}$ be the function defined by

$$f(x) = \begin{cases} 0 & \text{if } |x| \geq 1, \\ \exp\left(\dfrac{c}{x^2 - 1}\right) & \text{if } |x| < 1. \end{cases}$$

Then $f \in \mathcal{D}(\mathbb{R})$. Obviously $supp \, f = [-1, 1]$. It is also clear that f admits derivatives of any order at points t such that $|t| \neq 1$. For $|t_0| = 1$, one has $\lim_{t \to t_0} f(t) = 0$ and $f(t_0) = 0$. So f is continuous at t_0. Moreover, one obtains by induction that for $|t| < 1$ and $k \in \mathbb{N}^*$,

$$f^{(k)}(t) = \frac{P_k(t) f(t)}{(t^2 - 1)^{2k}}$$

where P_k is a polynomial of degree $3k - 2$. Finally considering the left and the right derivatives and the fact that $f^{(k)}(t) \xrightarrow[|t| \to 1]{} 0$, one verifies that f admits also derivatives of any order at t_0 with $|t_0| = 1$.

Example 1.2. From Example 1.1, one obtains for any interval $[a, b]$, $a < b$, a function $g \in \mathcal{D}(\mathbb{R})$ such that *supp* $g = [a, b]$. Indeed, it is sufficient to take the composite of the function f with the function φ defined by

$$\varphi(t) = \frac{2}{b-a}t - \frac{b+a}{b-a}$$

from $[a, b]$ onto $[-1, 1]$. We have

$$g(t) = \begin{cases} 0 & \text{if } t \notin \,]a, b[\,, \\ \exp\left(\dfrac{c\,(b-a)^2}{4\,(t-a)\,(t-b)}\right) & \text{if } t \in \,]a, b[\,. \end{cases}$$

Example 1.3. The analogue of Example 1.1 in $\mathbb{R}^n$ is the following

$$h(x) = \begin{cases} 0 & \text{if } \|x\| \geq 1, \\ \exp\left(\dfrac{c}{\|x\|^2 - 1}\right) & \text{if } \|x\| < 1. \end{cases}$$

Clearly *supp* $h = \{x \in \mathbb{R}^n : \|x\| \leq 1\}$. Also h is obviously of class $\mathcal{C}^\infty$ at any point x such that $\|x\| \neq 1$. For $\|a\| = 1$, one has $\lim\limits_{x \to a} h(x) = 0 = h(a)$. Hence h is continuous at a. Again an induction argument as in Example 1.1 shows that the partial derivatives of any order exist and are continuous at a. So f is in $\mathcal{D}(\mathbb{R}^n)$.

Example 1.4. From Example 1.1, one obtains another function $g \in \mathcal{D}(\mathbb{R}^n)$ such that *supp* $g = \{x \in \mathbb{R}^n : \|x\| \leq 1\}$. Indeed, it suffices to take the composite of the function f with the function ψ defined by

$$\psi(x_1, \ldots, x_n) = x_1^2 + x_2^2 + \cdots + x_n^2 = \|x\|^2$$

which is, of course, of class $\mathcal{C}^\infty$ from $\mathbb{R}^n$ to $\mathbb{R}$. Thus, we have

$$g(x) = \begin{cases} 0 & \text{if } \|x\| \geq 1, \\ \exp\left(\dfrac{c}{\|x\|^4 - 1}\right) & \text{if } \|x\| < 1. \end{cases}$$

1.7 The Topology of $\mathcal{D}^m(\Omega)$

One has $\mathcal{D}^m(\Omega) \subset \mathcal{E}^m(\Omega)$, $m \in \overline{\mathbb{N}}$. Of course $\mathcal{D}^m(\Omega)$ is a vector subspace of $\mathcal{E}^m(\Omega)$ but as a matter of fact, it is not complete for the topology induced by $\mathcal{E}^m(\Omega)$. To see this, take $\Omega = \mathbb{R}$ and $\varphi \in \mathcal{D}^m(\mathbb{R})$, $\varphi \geq 0$ and $\varphi \neq 0$ with *supp* $\varphi = [0, 1]$. Consider the sequence $(f_n)_n$ defined by

$$f_n(x) = \sum_{p=1}^{n} 2^{-p} \varphi(x - p).$$

It is Cauchy in $\mathcal{D}^m(\mathbb{R})$ for the topology induced by $\mathcal{E}^m(\mathbb{R})$. But its limit in $\mathcal{E}^m(\mathbb{R})$ does not have a compact support. Indeed take x_0 such that $\varphi(x_0) > 0$. Then for any positive integer k,

$$f(x_0 + k) = \sum_{p=1}^{+\infty} 2^{-p} \varphi \left[(x_0 + k) - p \right] \geq 2^{-k} \varphi(x_0) > 0.$$

We will then look for another *l.c.s.* topology τ on $\mathcal{D}^m(\Omega)$ such that $(\mathcal{D}^m(\Omega), \tau)$ is sequentially complete. Notice that

$$\mathcal{D}^m(\Omega) = \bigcup_K \mathcal{D}_K^m(\Omega),$$

where K runs over the collection of compact subsets of Ω and look for the topology τ so that $\tau \mid_{\mathcal{D}_K^m(\Omega)} = \tau_K$, where τ_K is the topology of $\mathcal{D}_K^m(\Omega)$. By definition $V \in \tau$ means that V is an open set for the topology τ. We are then led to consider

$$\mathcal{B} = \{ W \subset \mathcal{D}^m(\Omega) : W \text{ is a disc and } W \cap \mathcal{D}_K^m(\Omega) \in \tau_K, \ \forall K \}.$$

In order to have $\mathcal{B}$ as a basis of 0-neighborhoods for τ, the following condition is necessary

$$\forall V, \ V \in \tau, \ V = \bigcup_{\varphi \in V} (\varphi + W_\varphi), \ W_\varphi \in \mathcal{B}.$$

We therefore consider the collection τ of all unions of subsets V of $\mathcal{D}^m(\Omega)$ of the form

$$\varphi + W, \ \varphi \in \mathcal{D}^m(\Omega), \ W \in \mathcal{B}.$$

Lemma 1.1. *Let $V \in \tau$. Then for any $\varphi \in V$, there is $W_\varphi \in \mathcal{B}$, such that $\varphi + W_\varphi \subset V$.*

Proof. If $\varphi \in V$, then there is $\psi \in \mathcal{D}^m(\Omega)$ and $W_\psi \in \mathcal{B}$ such that $\varphi \in \psi + W_\psi \subset V$. Let K be a compact subset of Ω such that $\varphi, \ \psi \in \mathcal{D}_K^m(\Omega)$. One has $\varphi - \psi \in \mathcal{D}_K^m(\Omega) \cap W_\psi$. So there is $U_\varphi \in \mathcal{V}_{\tau_K}(0)$ such that $\varphi - \psi + U_\varphi \subset \mathcal{D}_K^m(\Omega) \cap W_\psi$. Let $\alpha_\varphi > 0$ such that $\alpha_\varphi (\varphi - \psi) \subset U_\varphi$. So $(1 + \alpha_\varphi)(\varphi - \psi) \in W_\psi$ or yet $\varphi - \psi \in (1 - \delta_\varphi) W_\psi$, with $0 < \delta_\varphi < 1$. Therefore one has $\varphi - \psi + \delta_\varphi W_\psi \subset W_\psi$ with W_ψ convex. Whence $\varphi + \delta_\varphi W_\psi \subset V$, so that $\delta_\varphi W_\psi \in \mathcal{B}$. So we proved that

$$\tau = \{V \subset \mathcal{D}^m(\Omega) \ / \ \forall \varphi \in V, \ \exists W_\varphi \in \mathcal{B} : \varphi + W_\varphi \subset V\}.$$

This implies that τ is a topology on $\mathcal{D}^m(\Omega)$ for which $\mathcal{B}$ is a 0-basis of neighborhoods. $\qquad\square$

Proposition 1.6. *The topology τ on $\mathcal{D}^m(\Omega)$ is locally convex. It is called the topology of $\mathcal{D}^m(\Omega)$.*

Proof. The origin admits a basis of absolutely convex neighborhoods. It is then sufficient to show that the addition and multiplication by scalars are continuous. The first assertion follows from the fact that for any $W \in \mathcal{B}$,

$$(\varphi + \frac{1}{2}W) + (\psi + \frac{1}{2}W) = \varphi + \psi + W; \ \forall \varphi, \ \psi \in \mathcal{D}^m(\Omega).$$

For the second, one writes

$$\alpha\varphi - \alpha_0\varphi_0 = \alpha(\varphi - \varphi_0) + (\alpha - \alpha_0)\varphi_0.$$

Take $\delta > 0$ such that $\delta\varphi_o \in \frac{1}{2}W$ and let c be such that

$$2(|\alpha_0| + \delta)\,|c| \leq 1.$$

Then $\alpha\varphi - \alpha_0\varphi_0 \in W$, whenever $|\alpha - \alpha_0| < \delta$ and $\varphi - \varphi_0 \in c\,W$. $\qquad\square$

Remark 1.5.

(1) Notice that for any compact subset K of Ω one has

$$\forall V \in \tau, \ V \cap \mathcal{D}_K^m(\Omega) \in \tau_K.$$

Indeed if $\varphi \in V \cap \mathcal{D}_K^m(\Omega)$ then, by the construction of τ, there is $W_\varphi \in \mathcal{B}$ such that

$$(\varphi + W_\varphi) \cap \mathcal{D}_K^m(\Omega) \subset V \cap \mathcal{D}_K^m(\Omega).$$

Since $W_\varphi \cap \mathcal{D}_K^m(\Omega)$ is a 0-neighborhood in $\mathcal{D}_K^m(\Omega)$, the set $V \cap \mathcal{D}_K^m(\Omega)$ is a neighborhood of φ in $\mathcal{D}_K^m(\Omega)$. So $V \cap \mathcal{D}_K^m(\Omega) \in \tau_K$.

(2) Let W be a balanced and convex subset of $\mathcal{D}^m(\Omega)$. It is immediate, by (1), that W is in τ if and only if it is in $\mathcal{B}$.

We come now to fundamental properties of the topology τ.

Proposition 1.7. *For any compact subset K of Ω, one has $\tau_{|_{\mathcal{D}_K^m(\Omega)}} = \tau_K$.*

Proof. Let K be a given compact subset of Ω. By (1) of Remark 1.5, $\tau|_{\mathcal{D}_K^m(\Omega)}$ is coarser than τ_K. Conversely, for $U \in \tau_K$ let us show that there is $V \in \tau$ such that $U = V \cap \mathcal{D}_K^m(\Omega)$. We know that, for any $\varphi \in U$, there is N_φ and $\varepsilon > 0$ such that

$$B_{K,N_\varphi}(\varphi, \varepsilon) = \left\{ \psi \in \mathcal{D}_K^m(\Omega) : p_{K,N_\varphi}(\psi - \varphi) < \varepsilon \right\} \subset U.$$

But

$$B_{K,N_\varphi}(\varphi, \varepsilon) = \varphi + \left\{ \chi \in \mathcal{D}_K^m(\Omega) : p_{K,N_\varphi}(\chi) < \varepsilon \right\}$$

$$= \varphi + \left\{ \chi \in \mathcal{D}_K^m(\Omega) : \max_{|j| \leq N_\varphi} \sup_{x \in K} \left| \chi^{(j)}(x) \right| < \varepsilon \right\} \cap \mathcal{D}_K^m(\Omega)$$

$$= \varphi + W_\varphi \cap \mathcal{D}_K^m(\Omega),$$

where

$$W_\varphi = \left\{ \chi \in \mathcal{D}^m(\Omega) : \max_{|j| \leq N_\varphi} \sup_{x \in \Omega} \left| \chi^{(j)}(x) \right| < \varepsilon \right\}.$$

It is clear that $W_\varphi \in \mathcal{B}$. Then $V = \bigcup_{\varphi \in U} (\varphi + W_\varphi)$ answers the question. $\quad\square$

Proposition 1.8. *The space* $(\mathcal{D}^m(\Omega), \tau)$ *is sequentially complete.*

Proof. We will show that any Cauchy sequence in $\mathcal{D}^m(\Omega)$ is contained in some $\mathcal{D}_K^m(\Omega)$. It will be Cauchy there by the previous proposition. The conclusion then follows from the completeness of $\mathcal{D}_K^m(\Omega)$ and again from the previous proposition. Take $B = (f_n)_n$ a Cauchy sequence in $\mathcal{D}^m(\Omega)$ and let $(K_q)_q$ be an exhaustive sequence of compact subsets of Ω. If B were not contained in some $\mathcal{D}_K^m(\Omega)$, there would exist a sequence $(x_q)_q$ in Ω and a subsequence $(f_{n_q})_q$ of B, denoted by $(\varphi_q)_q$, such that $\varphi_q(x_q) \neq 0$ (Take $\varphi_q \notin \mathcal{D}_{K_q}^m(\Omega)$ and $x_q \notin K_q$). We will exhibit a 0-neighborhood W in $\mathcal{D}^m(\Omega)$ which does not absorb B. For that it suffices to find W such that $\varphi_q \notin qW$, for every q. Since the exhaustive sequence is increasing, any compact subset of Ω contains only a finite number of elements of the sequence $(x_q)_q$, for it is contained in some K_q. Given K a compact subset, let x_{q_i}, $i = 1, ..., l$, be the elements of the sequence that it contains. Put

$$W_{K,\varepsilon} = \{ \psi \in \mathcal{D}_K^m(\Omega) : \ p_{K,0}(\psi) < \varepsilon \}.$$

For $\varphi_{q_i} \notin q_i W_{K,\varepsilon}$, it suffices that $\varepsilon \leq \frac{1}{q_i} |\varphi_{q_i}(x_{q_i})|$. Take

$$W_K = \left\{ \psi \in \mathcal{D}_K^m(\Omega) : \ p_{K,0}(\psi) < \frac{1}{q_i} |\varphi_{q_i}(x_{q_i})|, \quad i = 1, ..., l \right\}.$$

Notice that

$$W_K = \left\{ \psi \in \mathcal{D}_K^m(\Omega) : \ p_{K,o}(\psi) < \frac{1}{q} |\varphi_q(x_q)| ; \ \forall q \right\}$$

$$= \left\{ \psi \in \mathcal{D}^m(\Omega) : \ \sup_{x \in \Omega} |\psi(x)| < \frac{1}{q} |\varphi_q(x_q)| ; \ \forall_q \right\} \cap \mathcal{D}_K^m(\Omega).$$

Then

$$W = \left\{ \psi \in \mathcal{D}^m(\Omega) : \ \sup_{x \in \Omega} |\psi(x)| < \frac{1}{q} |\varphi_q(x_q)| ; \ \forall_q \right\}$$

is the neighborhood looked for. □

Remark 1.6. The previous presentation is for the convenience of the reader. Actually (cf. [1], E.V.T. II 29, proposition 5), $\mathcal{B}$ is a fundamental system of neighborhoods of zero, in $\mathcal{D}^m(\Omega)$, for a locally convex topology τ. The latter is nothing else than the collection of unions of subsets V of $\mathcal{D}^m(\Omega)$ of the form

$$\varphi + W, \ \varphi \in \mathcal{D}^m(\Omega), \ W \in \mathcal{B}.$$

Also τ induces on each $\mathcal{D}_K^m(\Omega)$ its own topology (cf. [1], E.V.T. II 35, proposition 9).

As a consequence, we have the following useful result.

Proposition 1.9. *A sequence* $(f_n)_n$ *in* $\mathcal{D}^m(\Omega)$ *converges to an element* f *of* $\mathcal{D}^m(\Omega)$ *if and only if*

(1) there is a compact subset K *of* Ω *such that* supp $f \subset K$ *and* supp $f_n \subset K$, *for every* n.
(2) $(f_n)_n$ *converges to* f *in* $\mathcal{D}_K^m(\Omega)$.

Here is by the way a very useful fundamental property of the topology of $\mathcal{D}^m(\Omega)$.

Proposition 1.10. *A mapping* f *from* $\mathcal{D}^m(\Omega)$ *into any topological space is continuous if and only if its restriction* f_K *to every* $\mathcal{D}_K^m(\Omega)$ *is continuous.*

Remark 1.7. Since the spaces $\mathcal{D}_K^m(\Omega)$ are metrizable, the continuity of a mapping f from $\mathcal{D}^m(\Omega)$ into any topological space can be expressed via sequences.

1.8 Density results

One has

$$\mathcal{D}(\Omega) \subset \mathcal{D}^m(\Omega) \subset \mathcal{E}^m(\Omega), \text{ for every } m \in \mathbb{N}.$$

We will show that

(1) $\mathcal{D}(\Omega)$ is dense in $\mathcal{D}^m(\Omega)$, for every $m \in \mathbb{N}$.
(2) $\mathcal{D}^m(\Omega)$ is dense in $\mathcal{E}^m(\Omega)$, for every $m \in \mathbb{N}$.

For the first result, we will use the so-called regularization method.

Definition 1.1. A sequence $(\theta_j)_{j\in\mathbb{N}}$ in $\mathcal{D}(\mathbb{R}^n)$ is said to be regularizing if it satisfies the following conditions

(1) $\theta_j(x) \geq 0$, for every $x \in \mathbb{R}^n$.
(2) $supp\ \theta_j \subset \overline{B}(0, \varepsilon_j)$, with $(\varepsilon_j)_j$ tending to 0.
(3) $\displaystyle\int_{\mathbb{R}^n} \theta_j(x)dx = 1$.

Such sequences do always exist. Indeed it suffices to take $\varepsilon_j = (j+1)^{-1}$, $\theta_j(x) = (j+1)^n \theta[(j+1)x]$, where θ is in $\mathcal{D}(\mathbb{R}^n)$ with $supp\ \theta \subset \overline{B}(0,1)$ and $\displaystyle\int_{\mathbb{R}^n} \theta(x)dx = 1$.

Let $(\theta_j)_j$ be a regularizing sequence and $f \in \mathcal{E}^m(\mathbb{R}^n)$, $m \in \overline{\mathbb{N}}$. The functions $f_j = f * \theta_j$ the convolutions of f with the θ_j's are said to be the regularizations of f. They are indeed of class C^∞. Notice that

$$f_j(x) - f(x) = \int_{\|y\|\leq\varepsilon_j} [f(x-y) - f(x)]\,\theta_j(y)dy.$$

Due to this formula, the functions f_j play an essential role in density results.

Theorem 1.1 (Regularization theorem). *Let $(\theta_j)_j$ be a regularizing sequence and X any one of the spaces $\mathcal{D}^m(\mathbb{R}^n)$, $\mathcal{K}(\mathbb{R}^n)$, $\mathcal{E}^m(\mathbb{R}^n)$, or $\mathcal{L}^p(\mathbb{R}^n)$. Then for every $f \in X$, one has $f = \lim_{j\to+\infty} f_j$ in X.*

Proof. Notice first that for every $j \in \mathbb{N}$, f_j is of class C^∞ on $\mathbb{R}^n$ and that

$$supp\ f_j \subset supp\ f + supp\ \theta_j.$$

(1) $X = \mathcal{C}(\mathbb{R}^n) = \mathcal{E}^0(\mathbb{R}^n)$. Let $f \in X$ and K be any compact subset of $\mathbb{R}^n$. One has

$$\sup_{x \in K} |f_j(x) - f(x)| \leq \int_{\|y\| \leq \varepsilon_j} \sup_{x \in K} |f(x - y) - f(x)| \, \theta_j(y) dy$$
$$\leq \sup_{\substack{x \in K \\ \|y\| \leq \varepsilon_j}} |f(x - y) - f(x)| .$$

Notice that if $x \in K$ and $y \in \overline{B}(0, \varepsilon_j)$, then

$$x - y \in K + \overline{B}(0, \varepsilon_j) \text{ and } x \in K + \overline{B}(0, \varepsilon_j).$$

We can suppose $\varepsilon_j \leq 1$ and therefore

$$K + \overline{B}(0, \varepsilon_j) \subset K + \overline{B}(0, 1), \text{ for every } j.$$

The uniform continuity of f on the compact set $K_1 = K + \overline{B}(0, 1)$ allows the conclusion.

(2) $X = \mathcal{E}^m(\mathbb{R}^n) = \mathcal{C}^m(\mathbb{R}^n)$. Let $f \in \mathcal{E}^m(\mathbb{R}^n)$. For any multi-index l with $|l| \leq m$, one has the following derivation formula

$$D^l f_j = (D^l f) * \theta_j = \theta_j * D^l f.$$

The result follows then from (1).

(3) $X = \mathcal{K}(\mathbb{R}^n) = \mathcal{D}^0(\mathbb{R}^n)$. Let $f \in \mathcal{K}(\mathbb{R}^n)$. With $\varepsilon_j \leq 1$, the f_j's and f have their supports in the compact subset $K = supp\, f + \overline{B}(0, 1)$; then apply (1).

(4) $X = \mathcal{D}^m(\mathbb{R}^n)$. As in (2), using (3) instead of (1).

(5) Let $f \in \mathcal{L}^p(\mathbb{R}^n)$ with $1 \leq p < +\infty$. One has

$$f_j \in \mathcal{L}^p(\mathbb{R}^n) \text{ and } \|f_j\|_p \leq \|\theta_j\|_1 \|f\|_p = \|f\|_p$$

since $\|\theta_j\|_1 = 1$, for every $j \in \mathbb{N}$. But $\mathcal{K}(\mathbb{R}^n)$ is dense in $\mathcal{L}^p(\mathbb{R}^n)$. Hence for any $\varepsilon > 0$, there is $\varphi \in \mathcal{K}(\mathbb{R}^n)$ such that $\|f - \varphi\|_p \leq \varepsilon$. Then $\|f_j - \varphi_j\|_p \leq \|f - \varphi\|_p \leq \varepsilon$ and so

$$\|f - f_j\|_p \leq \|f - \varphi\|_p + \|\varphi - \varphi_j\|_p + \|\varphi_j - f_j\|_p$$
$$\leq 2\varepsilon + \|\varphi - \varphi_j\|_p .$$

Now when $j \to +\infty$, the sequence $(\varphi_j)_j$ converges to φ in $\mathcal{K}(\mathbb{R}^n)$, by (3); hence $\|\varphi - \varphi_j\|_p$ converges to 0. Whence $\|\varphi - \varphi_j\|_p \leq \varepsilon$, for j large enough, so that $\|f - f_j\|_p \leq 3\varepsilon$, for j large enough. $\qquad\qquad\square$

Theorem 1.2. $\mathcal{D}(\Omega)$ *is dense in* $\mathcal{D}^m(\Omega)$, *for every* $m \in \mathbb{N}$.

Proof. If $\Omega = \mathbb{R}^n$, the result follows from the regularization theorem. If $\Omega \neq \mathbb{R}^n$, let $f \in \mathcal{D}^m(\Omega)$ with support K. Extending f by 0 outside of Ω, one obtains a function $g \in \mathcal{D}^m(\mathbb{R}^n)$ the restriction of which to Ω is equal to f. Putting $g_j = g * \theta_j$, one has

$$supp\, g_j \subset K + \overline{B}(0, \varepsilon_j).$$

To have $supp\, g_j \subset \Omega$, just take $\varepsilon_j < d(K, \Omega^c)$. Supposing $(\varepsilon_j)_j$ decreasing, one has $K + \overline{B}(0, \varepsilon_j) \subset K + \overline{B}(0, \varepsilon_1)$. The g_j's and g have their supports in the compact subset $K_1 = K + \overline{B}(0, \varepsilon_1)$ of Ω. Consider the restrictions φ_j of g_j to Ω. They are in $\mathcal{D}^m(\Omega)$. Moreover for every multi-index l with $|l| \leq m$, one has

$$\sup_{x \in K_1} \left| D^l f(x) - D^l \varphi_j(x) \right| = \sup_{x \in K_1} \left| D^l g(x) - D^l g_j(x) \right|.$$

Finally, by the regularization theorem, the right-hand side tends to 0. $\square$

We will show that $\mathcal{D}^m(\Omega)$ is dense in $\mathcal{E}^m(\Omega)$, for every $m \in \overline{\mathbb{N}}$. For any $f \in \mathcal{E}^m(\Omega)$, we have to find functions with compact supports approaching f. Multiplication by characteristic functions of compact subsets does not preserve derivability (in fact, even continuity). It is the technique called of truncating that solves the question.

Definition 1.2. A truncating sequence on Ω, associated with an exhaustive sequence $(K_j)_j$ of compact subsets of Ω, is any sequence $(\psi_j)_j \subset \mathcal{D}(\Omega)$ of functions such that

(1) $0 \leq \psi_j \leq 1$,
(2) $\psi_j = 1$ on K_j,
(3) $supp\, \psi_j \subset \overset{o}{K}_{j+1}$.

Such sequences do exist. Indeed the separation lemma of Urysohn (cf. [15]) gives, for every j, a function $\psi_j \in \mathcal{D}(\overset{o}{K}_{j+1})$ such that $0 \leq \psi_j \leq 1$, $\psi_j = 1$ on K_j and $supp\, \psi_j \subset \overset{o}{K}_{j+1}$. We extend each ψ_j to Ω by 0 outside of $\overset{o}{K}_{j+1}$.

Theorem 1.3. $\mathcal{D}^m(\Omega)$ *is dense in* $\mathcal{E}^m(\Omega)$, *for every* $m \in \overline{\mathbb{N}}$.

Proof. Let $f \in \mathcal{E}^m(\Omega)$ and $(\psi_j)_j$ be a truncating sequence on Ω. The sequence $(\psi_j f)_j$ is in $\mathcal{D}^m(\Omega)$. Moreover, on any compact subset K one has $\psi_j f = f$, for j large enough. So the sequence $(\psi_j f)_j$ converges uniformly to f in any compact subset; and similarly $D^l(\psi_j f)$ to $D^l f$ for every multi-index l with $|l| \leq m$. $\square$

Combining the two previous density theorems, one obtains the following result.

Theorem 1.4. $\mathcal{D}(\Omega)$ *is dense in* $\mathcal{E}^m(\Omega)$, *for every* $m \in \overline{\mathbb{N}}$.

Spaces of integrable functions play an important role in the theory of distributions. Notice that

$$\mathcal{D}(\Omega) \subset \mathcal{D}^m(\Omega) \subset \mathcal{K}(\Omega) \subset \mathcal{L}^p(\Omega), \ 1 \le p < +\infty.$$

So a density result is in order. We know that $\mathcal{K}(\Omega) = \mathcal{D}^0(\Omega)$ is dense in $L^p(\Omega)$, $1 \le p < +\infty$ and that $\mathcal{D}(\Omega)$ is dense in $\mathcal{K}(\Omega)$.

Proposition 1.11. $\mathcal{D}(\Omega)$ *is dense in* $L^p(\Omega)$, $1 \le p < +\infty$.

We finish this chapter by the following recapitulating diagram:

$$\begin{array}{ccccccc} \mathcal{D}(\Omega) & \subset & \mathcal{D}^m(\Omega) & \subset & \mathcal{K}(\Omega) & \subset & L^p(\Omega) \\ \cap & & \cap & & \cap & & \cap \\ \mathcal{E}(\Omega) & \subset & \mathcal{E}^m(\Omega) & \subset & \mathcal{C}(\Omega) & \subset & L^p_{loc}(\Omega) \end{array}$$

where $1 \le p \le +\infty$.

The space $\mathcal{L}^1_{loc}(\Omega)$ (of locally integrable functions on Ω) contains all spaces of functions considered in this chapter. We will see that it plays a particular role in the theory of distributions.

The very basic spaces of Distribution Theory (fundamental spaces or test function spaces) are fully described. Their topologies are given in detail. These spaces often appear in several branches of mathematics. So this chapter may be useful for people who are not necessarily working in distribution theory. The notations are introduced step by step and remarks indicate the specific ones to the theory. Observe that Section 1.7 on the topology of $\mathcal{D}^m(\Omega)$ has to be carefully read, it is essential for the rest of the book. The chapter ends with density results which are frequently used in the sequel.

Among the useful inclusions one has

$$\mathcal{D}(\Omega) \subset \mathcal{D}^m(\Omega) \subset \mathcal{E}^m(\Omega), \text{ for every } m \in \mathbb{N}.$$

One shows that $\mathcal{D}(\Omega)$ is dense in $\mathcal{D}^m(\Omega)$ using the regularization method (see Theorem 1.1, p. 11). To obtain the density of $\mathcal{D}^m(\Omega)$ in $\mathcal{E}^m(\Omega)$, one needs the truncating method; by combining the two methods one gets the density of the first space in the third. Notice that all spaces of functions, considered in this chapter are subspaces of $\mathcal{L}^1_{loc}(\Omega)$. The latter actually plays a special role in distribution theory.

The subject matter of the chapter is contained in all books dealing with distributions. In our presentation, there are no long preliminaries on locally convex spaces nor on inductive limits as for instance in [15]. We have tried to make the proofs free of "tricks" and non necessary generalities. For more readings, see [3], [4], [5], [6], [7], [8], [12], [14], [15], and [16].

Chapter 2

Distributions

This chapter deals with general properties of distributions. The first classical examples are presented (regular distributions, measures, principal value of Cauchy, finite parts of Hadamard,...). Notice that $\mathcal{L}^1_{loc}(\Omega)$ is embedded in $\mathcal{D}'(\Omega)$ the set of distributions on the open subset Ω of $\mathbb{R}^n$. The notion of a support of a distribution is discussed, as well as the class of distributions with compact support. Endowed with the usual addition and multiplication by scalars, $\mathcal{D}'(\Omega)$ is a vector space. We also consider the multiplication of a distribution by a function and the derivability of a distribution. The space $\mathcal{D}'(\Omega)$ is endowed with a topology called vague. The linear differential operators with C^∞ coefficients on Ω, from $\mathcal{D}'(\Omega)$ into $\mathcal{D}'(\Omega)$, are then continuous.

2.1 Definitions and characterizations

Definition 2.1. A distribution on Ω is any continuous linear form on $\mathcal{D}(\Omega)$. The set of distributions on Ω is denoted $\mathcal{D}'(\Omega)$.

Here are classical characterizations which are very useful in practice.

Proposition 2.1. Let $T : \mathcal{D}(\Omega) \longrightarrow \mathbb{C}$ be a linear form. The following assertions are equivalent.

(1) T is a distribution on Ω.
(2) For any sequence $(\varphi_n)_n$ converging to zero in $\mathcal{D}(\Omega)$, the sequence $(T(\varphi_n))_n$ converges to zero in $\mathbb{C}$.
(3) For any compact subset K in Ω, there is an integer $N \geq 0$ and a constant $c > 0$ such that
$$|T(\varphi)| \leq c\, p_{K,N}(\varphi), \ \forall \varphi \in \mathcal{D}_K(\Omega).$$

Proof.   Properties (2) and (3) express in two different manners the fact that the restriction of T to each $\mathcal{D}_K(\Omega)$ is continuous; that is a characterization of the continuity of T by Proposition 1.9 of Chapter 1. Now (2) comes from the fact that the spaces $D_K(\Omega)$ are metrizable. For (3), the continuity of T at the origin is written

$$\forall \varepsilon > 0,\ \exists \eta > 0,\ \exists N \in \mathbb{N}^* : p_{K,N}(\varphi) \leq \eta \Longrightarrow |T(\varphi)| \leq \varepsilon.$$

For any $\alpha > 0$, one has

$$p_{K,N}\left(\frac{\eta\varphi}{p_{K,N}(\varphi)+\alpha}\right) \leq \eta,\ \forall \varphi \in \mathcal{D}_K(\Omega).$$

Whence

$$|T(\varphi)| \leq \frac{\varepsilon}{\eta}\left[p_{K,N}(\varphi)+\alpha\right],\ \forall \varphi \in \mathcal{D}_K(\Omega)$$

and this for any $\alpha > 0$. Hence

$$|T(\varphi)| \leq c\, p_{K,N}(\varphi),\ \forall \varphi \in \mathcal{D}_K(\Omega)\,;\text{ with } c = \frac{\varepsilon}{\eta}. \qquad \square$$

The integer N of (3), in the previous proposition, depends a priori on the compact set K. But it may not depend of it. In this sense we have the following definition.

Definition 2.2. A distribution T is said to be of order less than or equal to a given positive integer m if,

$$\forall K,\ \exists c > 0 : |T(\varphi)| \leq c\, p_{K,m}(\varphi),\ \forall \varphi \in \mathcal{D}_K(\Omega).$$

We have the following characterization.

Proposition 2.2. *Let* $T : \mathcal{D}(\Omega) \longrightarrow \mathbb{C}$ *be a linear form. The following assertions are equivalent.*

(1) T is a distribution of order $\leq m$.
(2) T is continuous on $\mathcal{D}(\Omega)$ for the topology of $\mathcal{D}^m(\Omega)$.
(3) T extends to a continuous linear form on $\mathcal{D}^m(\Omega)$.

Proof.   (1) $\Longrightarrow$ (2) This ensues just from the definition.
(2) $\Longrightarrow$ (3) Comes from the fact that $\mathcal{D}(\Omega)$ is dense in $\mathcal{D}^m(\Omega)$.
(3) $\Longrightarrow$ (1) Let $\widetilde{T} : \mathcal{D}^m(\Omega) \longrightarrow \mathbb{C}$ be the extension of T to $\mathcal{D}^m(\Omega)$. It suffices then to express the continuity of the restriction T, of $\widetilde{T}$ to $\mathcal{D}(\Omega)$, for the induced topology. $\qquad \square$

Notation 2.1. The value $T(\varphi)$ is also written $\langle T, \varphi \rangle$.

2.2 Examples

2.2.1. Locally integrable functions

For $f \in \mathcal{L}^1_{loc}(\Omega)$, one defines T_f by

$$\langle T_f, \varphi \rangle = \int_\Omega f(x)\varphi(x)dx, \ \varphi \in \mathcal{D}(\Omega).$$

It is linear. Moreover if *supp* $\varphi \subset K$, one has

$$|\langle T_f, \varphi \rangle| \leq \sup_{x \in K} |\varphi(x)| \int_K |f(x)|\, dx \leq c_K \, p_{K,0}(\varphi).$$

So T_f is a distribution of order zero.

2.2.2. Measures

Recall that a Radon measure μ on Ω is a continuous linear form on $\mathcal{K}(\Omega)$, i.e.,

$$\forall K, \ \exists c_K > 0 : |\langle \mu, \varphi \rangle| \leq c_K \, p_{K,0}(\varphi), \ \forall \varphi \in \mathcal{K}_K(\Omega).$$

Hence the restriction of μ to $\mathcal{D}(\Omega)$ is a distribution of order zero. In fact by Proposition 2.2, we have the following.

Proposition 2.3. *Let* $T : \mathcal{D}(\Omega) \longrightarrow \mathbb{C}$ *be a linear form. The following assertions are equivalent.*

(1) T is a distribution of order zero.
(2) T is continuous on $\mathcal{D}(\Omega)$ for the topology of $\mathcal{K}(\Omega)$.
(3) T extends to a Radon measure on Ω.

2.2.3. Dirac Measures

For $a \in \Omega$, one defines $\delta_a : \mathcal{D}(\Omega) \longrightarrow \mathbb{C}$ by $\langle \delta_a, \varphi \rangle = \varphi(a)$. It is a distribution of order zero. It extends to a Radon measure on Ω, still denoted δ_a. It is called a Dirac measure. It is not of the form T_f of the first example. Indeed suppose there is $f \in \mathcal{L}^1_{loc}(\mathbb{R})$ such that $\delta_0 = T_f$, i.e.,

$$\varphi(0) = \int_\mathbb{R} f(x)\varphi(x)dx, \ \forall \varphi \in \mathcal{D}(\mathbb{R}).$$

Taking $\varphi_n(x) = \theta\,[(n+1)x]$, where

$$\theta(x) = \begin{cases} 0 & \text{if } |x| \geq 1 \\ \exp\dfrac{1}{x^2-1} & \text{if } |x| < 1, \end{cases}$$

one obtains

$$\frac{1}{e} = \int_{\mathbb{R}} f(x)\varphi_n(x)dx, \ \forall n \in N.$$

But by the Lebesgue theorem the right-hand side tends to zero; whence an absurdity. In the sequel δ_0 is simply denoted δ.

A Radon measure is said to be regular if it is of the form T_f with $f \in \mathcal{L}_{loc}^1(\Omega)$. Otherwise it is said to be singular.

2.2.4. Cauchy principal value

The function $f : x \longmapsto \dfrac{1}{x}$ is not locally integrable on $\mathbb{R}$ (Take the interval $[-1, 1]$). It is locally integrable on $\mathbb{R}^* = \mathbb{R}\backslash\{0\}$. Hence it defines a distribution T_f of order zero on $\mathbb{R}^*$. It is natural to wonder if there is a distribution T on $\mathbb{R}$ extending T_f. For every $\varphi \in \mathcal{D}(\mathbb{R})$ and any $\alpha, \beta > 0$, put

$$I_{\alpha,\beta} = \int_{-\infty}^{-\alpha} \frac{\varphi(x)}{x}dx + \int_{\beta}^{+\infty} \frac{\varphi(x)}{x}dx$$

which is well defined. It is known that

$$\varphi(x) = \varphi(0) + x\psi(x); \ \psi \in \mathcal{C}(\mathbb{R})$$

with

$$\sup_{x \in supp \ \varphi} |\psi(x)| \le c \sup_{x \in supp \ \varphi} |\varphi'(x)|, \text{ where } c \text{ is a positive constant.}$$

One has

$$I_{\alpha,\beta} = \int_{-\infty}^{-\alpha} \psi(x)dx + \int_{\beta}^{+\infty} \psi(x)dx + \varphi(0)\left[\ln \alpha - \ln \beta\right].$$

This quantity does not in general admit a limit when α and β tend to zero. However if $\alpha = \beta$,

$$\lim_{\alpha \to 0} I_{\alpha,\alpha} = \int_{-\infty}^{+\infty} \psi(x)dx.$$

We define this way a distribution on $\mathbb{R}$ of order less or equal to 1, which extends T_f. It is denoted $V_p\dfrac{1}{x}$ and called the Cauchy principal value of $\dfrac{1}{x}$. It is given by

$$\left\langle V_p\frac{1}{x}, \varphi \right\rangle = \lim_{\varepsilon \to 0} \int_{|x| \ge \varepsilon} \frac{\varphi(x)}{x}dx; \ \varphi \in \mathcal{D}(\mathbb{R}).$$

2.2.5. Hadamard finite parts

The function $f : x \longmapsto \dfrac{1}{x^2}$ is not locally integrable on $\mathbb{R}$. But it is so on $\mathbb{R}^*$. If we want to proceed as in **2.2.4**, let $\varphi \in \mathcal{D}(\mathbb{R})$ and $A > 0$ such that $supp\, \varphi \subset [-A, A]$. One has

$$\varphi(x) = \varphi(0) + x\varphi'(0) + x^2\psi(x); \ \psi \in \mathcal{C}(\mathbb{R})$$

with

$$\sup_{|x|\leq A} |\psi(x)| \leq c \sup_{|x|\leq A} |\varphi''(x)|, \text{where } c \text{ is a positive constant.}$$

Then for any $\varepsilon > 0$,

$$\int_{|x|\geq\varepsilon} \frac{\varphi(x)}{x^2} dx = 2\frac{\varphi(0)}{\varepsilon} + 2\frac{\varphi(0)}{A} + \int_{|x|\geq\varepsilon} \psi(x)dx.$$

In general this quantity does not admit a limit when ε tends to zero. However

$$\lim_{\varepsilon\to 0} \left[\int_{|x|\geq\varepsilon} \frac{\varphi(x)}{x^2} dx - 2\frac{\varphi(0)}{\varepsilon} \right] = \frac{2\varphi(0)}{A} + \int_{-A}^{A} \psi(x)dx.$$

We define this way a distribution on $\mathbb{R}$ of order less or equal to 2, which extends T_f. It is called the finite part of $\dfrac{1}{x^2}$ and is denoted by $Pf\dfrac{1}{x^2}$. It is given by

$$\left\langle Pf\frac{1}{x^2}, \varphi \right\rangle = \lim_{\varepsilon\to 0} \left[\int_{|x|\geq\varepsilon} \frac{\varphi(x)}{x^2} dx - 2\frac{\varphi(0)}{\varepsilon} \right]; \ \varphi \in \mathcal{D}(\mathbb{R}).$$

Remark 2.1.

(1) In the same way, one defines $Pf\dfrac{1}{x^k}$ for any integer $k \geq 3$.

(2) The previous procedure is not valid for every locally integrable function on $\mathbb{R}^*$. An example is the function $x \longmapsto \exp\dfrac{1}{x^2}$.

2.2.6. Distributions of infinite order

For any $\varphi \in \mathcal{D}(\mathbb{R})$, put

$$\langle T, \varphi \rangle = \lim_{n\to+\infty} \sum_{p=0}^{n} \varphi^{(p)}(p).$$

This limit always exists for $\sum_{p=0}^{+\infty} \varphi^{(p)}(p)$ which in fact is a finite sum for any $\varphi \in \mathcal{D}(\mathbb{R})$. One verifies that T is a distribution. It is not of finite order.

To see this let us show that it is not of order less or equal to $j-1$ for each $j \geq 1$. Let $K = \left[-\dfrac{1}{2} + j, \dfrac{1}{2} + j \right]$ for $j \geq 1$. For any $\varphi \in \mathcal{D}_K(\mathbb{R})$ one has $\langle T, \varphi \rangle = \varphi^{(j)}(j)$. We will exhibit a sequence $(\psi_n)_n$ in $\mathcal{D}_K(\mathbb{R})$ such that

$$\psi_n \xrightarrow[n]{} 0 \text{ in } \mathcal{D}_K(\mathbb{R}) \text{ and } \psi_n^{(j)}(j) \text{ does not tend to zero.}$$

Replacing if needed ψ_n by φ_n such that $\varphi_n(x) = \psi_n(x + j)$, we may look for ψ_n in $\mathcal{D}_{\left[-\frac{1}{2}, \frac{1}{2} \right]}(\mathbb{R})$ with

$$\psi_n \xrightarrow[n]{} 0 \text{ in } \mathcal{D}^{j-1}_{\left[-\frac{1}{2}, \frac{1}{2} \right]}(\mathbb{R}) \text{ and } \psi_n^{(j)}(0) \text{ does not tend to zero.}$$

Let $\psi \in \mathcal{D}_{\left[-\frac{1}{2}, \frac{1}{2} \right]}(\mathbb{R})$ such that $\psi^{(j)}(0) > 0$. We look for ψ_n under the form

$$\psi_n(x) = n^\alpha \psi(nx), \text{ with } \alpha \in \mathbb{R}.$$

One has

$$\psi_n^{(j)}(0) = n^{\alpha+j} \psi^{(j)}(0), \text{ whenever } \alpha \geq -j.$$

On the other hand

$$\max_{m \leq j-1} \sup_{|x| \leq \frac{1}{2}} \left| \psi_n^{(m)}(x) \right| \xrightarrow[n]{} 0, \text{ whenever } \alpha < 1 - j.$$

We should have $-j \leq \alpha < 1 - j$ hence it suffices to take $\alpha = -j + \varepsilon$ with $0 \leq \varepsilon < 1$; and so for example

$$\psi_n(x) = n^{-j+\frac{1}{2}} \psi(nx).$$

2.3   Embedding of $\mathcal{L}^1_{loc}(\Omega)$ in $\mathcal{D}'(\Omega)$

It is clear that if $f, g \in \mathcal{L}^1_{loc}(\Omega)$ and $f = g$ almost everywhere (*a.e.* in brief), then $T_f = T_g$. The converse is also true.

Proposition 2.4. *Let $f, g \in \mathcal{L}^1_{loc}(\Omega)$. If $T_f = T_g$, Then $f = g$, a.e..*

Proof.   By linearity, we are led to

$$T_f = 0 \Longrightarrow f = 0, \text{ a.e..}$$

For that it is sufficient to show that

$$\int_K f(x)dx = 0, \text{ for every compact set } K \text{ in } \Omega.$$

For every integer $j \geq 1$, put $\Omega_j = \left\{ x \in \Omega : d(x, K) < \dfrac{1}{j} \right\}$. It is an open subset of Ω containing K. By a lemma of Urysohn type (cf. [15]), there is a sequence $(\varphi_j)_j$ in $\mathcal{D}(\Omega)$ such that $0 \leq \varphi_j \leq 1$, $\varphi_j = 1$ on K and $supp \, \varphi_j \subset \Omega_j$. We have by hypothesis

$$\int_\Omega f(x)\varphi_j(x)dx = 0, \text{ for every } j.$$

Then, due to the fact that $\varphi_j \xrightarrow{j} \chi_K$, *a.e.*, where χ_K is the characteristic function of K, the result follows from Lebesgue's theorem.

$\square$

Remark 2.2. We have then seen that $\mathcal{L}^1_{loc}(\Omega)$ is embedded in $\mathcal{D}'(\Omega)$.

2.4 Support of a distribution

Notice that for any function f, the complement $O_f = (supp \, f)^c$ in Ω of $supp \, f$ is the greatest open subset of Ω on which f vanishes. If $f \in \mathcal{L}^1_{loc}(\Omega)$, one has

$$\langle T_f, \varphi \rangle = \int_{supp \, f} f(x)\varphi(x)dx.$$

Hence $\langle T_f, \varphi \rangle = 0$ if $supp \, \varphi \subset O_f$. Notice that if f is continuous, then $U_f \subset O_f$ for any open subset U_f of Ω such that $\langle T_f, \varphi \rangle = 0$ for every φ with $supp \, \varphi \subset U_f$. For this, it suffices to show that

$$\{x : f(x) \neq 0\} \subset (U_f)^c.$$

Let $x_0 \in \Omega$ such that $f(x_0) \neq 0$. We may suppose $f(x_0) > 0$. Since f is continuous, there is a neighborhood $V(x_0)$ of x_0 such that $f(x) > 0$ for any $x \in V(x_0)$. Therefore $x_0 \notin U_f$, for otherwise take $\varphi \in \mathcal{D}(\Omega)$ with $\varphi \geq 0$ and $supp \, \varphi \subset U_f \cap V(x_0)$. So O_f appears to be the greatest open subset such that $\langle T_f, \varphi \rangle = 0$ for every φ with $supp \, \varphi \subset O_f$. This fact suggests the following notions.

Definition 2.3. Let T be a distribution on Ω.

(1) We say that T vanishes on an open subset U of Ω, if $\langle T, \varphi \rangle = 0$ for every $\varphi \in \mathcal{D}(\Omega)$ such that $supp \, \varphi \subset U$.
(2) The zero open set of T is the greatest open subset (in case of existence) of Ω on which T vanishes.

(3) The support of T is the complement in Ω of the zero open set.

Example 2.1.

(1) We have seen that if $f \in \mathcal{C}(\Omega)$, then

$$supp T_f = supp\, f.$$

(2) For any function $f \in \mathcal{L}^1_{loc}(\Omega)$, one has

$$supp\, T_f \subset supp\, f.$$

But we do not always have the equality. Indeed for the characteristic function χ_Q of rationals, one has $supp(\chi_Q) = \mathbb{R}$ and $supp(T_{\chi_Q}) = \emptyset$.

(3) For any $a \in \Omega$, one has

$$supp\, \delta_a = \{a\}.$$

Remark 2.3. Let $T \in \mathcal{D}'(\Omega)$ and $\varphi \in \mathcal{D}(\Omega)$. If $supp\, \varphi \cap supp\, T = \emptyset$, then $\langle T, \varphi \rangle = 0$. In particular, if φ vanishes on a neighborhood of $supp\, T$ then $\langle T, \varphi \rangle = 0$. But if φ vanishes only on $supp\, T$, we do not necessarily have $\langle T, \varphi \rangle = 0$. Indeed let T be the distribution defined on $\mathbb{R}$ by $\langle T, \varphi \rangle = \varphi'(0)$. Clearly $supp\, T = \{0\}$. Take then $\psi(x) = x\varphi(x)$, where $\varphi \in \mathcal{D}(\mathbb{R})$ with $\varphi(x) = 1$ on $[-1, 1]$. However if T is a distribution of order $\leq m$ with $m \in \overline{\mathbb{N}}$ and if $\varphi \in \mathcal{D}^m(\Omega)$ vanishes as well as its derivatives of order $\leq m$ on $supp\, T$, then $\langle T, \varphi \rangle = 0$ (see Proposition 7.6 of Chapter 7).

In the general case, the existence of the support of a distribution is due to the following result.

Proposition 2.5 (Localization principle). *Let $T \in \mathcal{D}'(\Omega)$ and $(U_i)_{i \in I}$ be a family of open subsets of Ω. If T vanishes on each U_i, then T vanishes on $\bigcup_i U_i$.*

Proof. Let $(f_i)_{i \in I}$ be a locally finite C^∞-partition of unity in $U = \bigcup_i U_i$, subordinated to the covering $(U_i)_i$. For φ in $\mathcal{D}(\Omega)$, put $\varphi_i = \varphi f_i$. Then $\varphi_i \in \mathcal{D}(U_i)$ and $\varphi = \sum_i \varphi_i$ on U. This sum is in fact finite for there is only a finite number of f_i's which do not vanish on $supp\, \varphi$. We have $\langle T, \varphi \rangle = \sum_i \langle T, \varphi_i \rangle = 0.$ $\qquad\square$

2.5   Pasting principle

Let $(\Omega_i)_i$ be a covering of Ω and a distribution T_i on Ω_i, for each i. The question is to know whether there is a distribution T on Ω extending every T_i. This is not always possible. Indeed take in $\mathbb{R}$, $\Omega_1 =] - \infty, 1[$, $\langle T_1, \varphi \rangle = \varphi(0)$, $\Omega_2 =] - 1, +\infty[$ and $\langle T_2, \varphi \rangle = \varphi'(0)$. Here $T_1 \neq T_2$ on $\Omega_1 \cap \Omega_2$. However, we do have the following general result called the pieces pasting principle.

Proposition 2.6. *Let $(\Omega_i)_{i \in I}$ be an open covering of Ω and $T_i \in \mathcal{D}'(\Omega_i)$ for every $i \in I$. If $T_i = T_j$ on $\Omega_i \cap \Omega_j$ for any $i, j \in I$ such that $\Omega_i \cap \Omega_j \neq \emptyset$, then*

(1) There is a unique distribution T on Ω the restriction of which to every Ω_i is exactly T_i.

(2) If moreover all the T_i's are of order $\leq m$, then it is also so for T.

Proof.

(1) Uniqueness results from the localization principle. For the existence, let $(f_i)_{i \in I}$ be a locally finite C^∞-partition of the unity in Ω subordinated to the covering $(\Omega_i)_{i \in I}$. For $\varphi \in \mathcal{D}(\Omega)$, put

$$\langle T, \varphi \rangle = \sum_i \langle T_i, f_i \varphi \rangle.$$

One thus defines a linear form T on $\mathcal{D}(\Omega)$. Let us show that it is continuous on $\mathcal{D}_K(\Omega)$ for any compact subset K of Ω. Notice that if $\varphi \in \mathcal{D}_K(\Omega)$, then only a finite number of indices intervene in the sum defining T. It is then sufficient to show that the function $\varphi \longmapsto \langle T_i, f_i \varphi \rangle$ is continuous on $\mathcal{D}_K(\Omega)$ for every $i \in I$. This follows from the continuity, by hypothesis, of T_i and the continuity of the map $\varphi \longmapsto f_i \varphi$, from $\mathcal{D}_K(\Omega)$ into $\mathcal{D}(\Omega)$, due to the Leibniz formula. On the other hand one easily checks that $T \mid_{\Omega_i} = T_i$.

(2) If every T_i is continuous on $\mathcal{D}(\Omega)$ for the topology of $\mathcal{D}^m(\Omega)$, we proceed as in (1) to verify that it is so for T.

$\square$

2.6   Distributions with compact support

We will see that the distributions with compact support have interesting particular properties. We have seen that if $f \in \mathcal{C}(\Omega)$ has a compact support K, then $supp\, T_f = supp\, f$ and

$$\langle T_f, \varphi \rangle = \int_K f(x)\varphi(x)dx, \ \forall \varphi \in \mathcal{D}(\Omega).$$

In fact this integral is well defined for any $\varphi \in \mathcal{C}(\Omega)$. And one has

$$|\langle T_f, \varphi \rangle| \le \left(\int_K |f(x)|\, dx \right) \sup_{x \in K} |\varphi(x)| \le c_K p_{K,0}(\varphi).$$

So T_f extends to a continuous linear form on $\mathcal{C}(\Omega)$.

Now let $T \in \mathcal{D}'(\Omega)$ be with a compact support. By a lemma of Urysohn type (cf. [15]), there is a function $\chi \in \mathcal{D}(\Omega)$ such that $\chi = 1$ on a neighborhood V of the support of T. As $\varphi - \chi\varphi = 0$ on V for any $\varphi \in \mathcal{D}(\Omega)$, one has

$$\langle T, \varphi \rangle = \langle T, \chi\varphi \rangle, \ \forall \varphi \in \mathcal{D}(\Omega).$$

Since T is continuous, one obtains for $K = supp\, \chi$

$$|\langle T, \varphi \rangle| \le c_K\, p_{K,m}(\varphi), \ \forall \varphi \in \mathcal{D}_K(\Omega),$$

where c_K is a positive constant and m a positive integer. For any compact subset L of Ω and any $\varphi \in \mathcal{D}_L(\Omega)$, one always has $supp\, \chi\varphi \subset supp\, \chi$, and

$$|\langle T_f, \varphi \rangle| \le c_K p_{K,m}(\chi\varphi) \le c_K p_{L,m}(\varphi), \ \forall \varphi \in \mathcal{D}_L(\Omega).$$

So m coming from $supp\, \chi$ does not depend of the compact set L. Hence we can state the following result.

Proposition 2.7. *Any distribution with compact support is of finite order.*

Remark 2.4. The converse is false. Take for example the distribution $V_p \dfrac{1}{x}$.

Here is a characterization of the distributions with compact support.

Proposition 2.8. *Let T be a distribution on Ω. It is with compact support if and only if there is a positive integer m such that T extends to a continuous linear form on $\mathcal{E}^m(\Omega)$.*

Proof. If T has a compact support, then it is of finite order less or equal to a given m. Hence it is extendible to a continuous linear form on $\mathcal{D}^m(\Omega)$, still denoted by T (Proposition 2.2). Let $\chi \in \mathcal{D}(\Omega)$ such that $\chi = 1$ on a neighborhood V of $supp\, T$. One defines a linear form on $\mathcal{E}^m(\Omega)$, by putting $\left\langle \widetilde{T}, \varphi \right\rangle = \langle T, \chi\varphi \rangle$. It is continuous as a composition of continuous maps. Finally for any $\varphi \in \mathcal{D}^m(\Omega)$, one has $\left\langle \widetilde{T}, \varphi \right\rangle = \langle T, \varphi \rangle$ for $\varphi - \chi\varphi = 0$ on V. Conversely, if T is extendible to a continuous linear form on $\mathcal{E}^m(\Omega)$, then there is a compact set L and a constant $c > 0$ such that

$$|\langle T, \varphi \rangle| \le c p_{L,m}(\varphi), \; \forall \varphi \in \mathcal{D}(\Omega).$$

Whence $\langle T, \varphi \rangle = 0$ for any φ such that $supp\, \varphi \subset L^c$. So $supp\, T \subset L$. $\qquad\square$

2.7 Operations on distributions

Endowed with the usual addition and multiplication by scalars, $(\mathcal{D}^m(\Omega))'$ is a complex vector space. We will study other operations on $(\mathcal{D}^m(\Omega))'$.

2.7.1. Product of a distribution by a function
Let $f \in \mathcal{L}^1_{loc}(\Omega)$ and $g \in \mathcal{C}(\Omega)$. It is natural to define gT_f by T_{gf}. But

$$\langle T_{gf}, \varphi \rangle = \langle T_f, g\varphi \rangle, \; \text{for every } \varphi \in \mathcal{K}(\Omega).$$

Hence $\langle gT_f, \varphi \rangle = \langle T_f, g\varphi \rangle$ for every $\varphi \in \mathcal{K}(\Omega)$. In the general case where $T \in (\mathcal{D}^m(\Omega))'$ and $g \in \mathcal{C}^m(\Omega)$ with $m \in \overline{\mathbb{N}}$, we put

$$\langle gT, \varphi \rangle = \langle T, g\varphi \rangle, \; \forall \varphi \in \mathcal{C}^m(\Omega).$$

Proposition 2.9. *If $T \in (\mathcal{D}^m(\Omega))'$ and $g \in \mathcal{C}^m(\Omega)$, then*

(i) $gT \in (\mathcal{D}^m(\Omega))'$.
(ii) $\{x : g(x) \ne 0\} \cap supp\, T \subset supp\, gT \subset supp\, g \cap supp\, T$.

Proof.

(i) It is clear that gT is linear. It is continuous as a composition of continuous maps.
(ii) We will argue by using complements. For the second inclusion, it suffices to show that $(supp\, g)^c$ and $(supp\, T)^c$ are zero open sets for gT. If $\varphi \in \mathcal{D}^m(\Omega)$ with $supp\, \varphi \subset (supp\, g)^c$, then $\langle T, g\varphi \rangle = 0$. If

$supp\ \varphi \subset (supp\ T)^c$, one also has $\langle T, g\varphi \rangle = 0$ for $supp\ g\varphi \subset supp\ \varphi$. For the other inclusion, let $x \notin supp\ gT$. If $g(x) = 0$, it is finished. Otherwise we have to show that $x \notin supp\ T$, i.e., there is an open subset V containing x such that $\langle T, \psi \rangle = 0$ for every $\psi \in \mathcal{D}^m(V)$. Since g is continuous, there is an open neighborhood U_1 of x such that $g(x) \neq 0$ for every $x \in U_1$. On the other hand $x \notin supp\ gT$, hence there is an open neighborhood U_2 of x such that $\langle T, g\varphi \rangle = 0$ for every φ with $supp\ \varphi \subset U_2$. Let now $\psi \in \mathcal{D}^m(U_1 \cap U_2)$ and put

$$\varphi_0(x) = \begin{cases} \dfrac{\psi(x)}{g(x)} & \text{if } x \in U_1 \cap U_2 \\ 0 & \text{elsewhere.} \end{cases}$$

Then $\varphi_0 \in \mathcal{D}^m(\Omega)$ and we have $\langle T, \psi \rangle = \langle T, g\varphi_0 \rangle = 0$.   □

We will treat a particular case which has interesting applications. In the sequel x will designate the function $t \longmapsto t$ from $\mathbb{R}$ into $\mathbb{R}$. We will consider the product xT of a distribution by x. For any $\varphi \in \mathcal{D}(\mathbb{R})$, one has $\langle xT, \varphi \rangle = \langle T, x\varphi \rangle$, hence we are led to study the map $\varphi \longmapsto x\varphi$.

Proposition 2.10. *1) The linear map $F : \varphi \longmapsto x\varphi$, from $\mathcal{D}(\mathbb{R})$ into $\mathcal{D}(\mathbb{R})$, is one-to-one and*

$$ImF = Ker\delta.$$

2) The linear map $G : T \longmapsto xT$, from $\mathcal{D}'(\mathbb{R})$ into $\mathcal{D}'(\mathbb{R})$, is onto and

$$KerG = \mathbb{C}\delta,$$

the complex straight line generated by δ.

Proof.   1) If $F(\varphi) = 0$, then $\varphi(t) = 0$ for every $t \neq 0$. Whence $\varphi = 0$ since it is continuous. On the other hand, it is clear that $Im\ F \subset Ker\delta$. Conversely, for $\varphi \in Ker\delta$, put

$$\psi(t) = \begin{cases} \dfrac{\varphi(t)}{t} & \text{if } t \neq 0 \\ \varphi'(0) & \text{if } t = 0. \end{cases}$$

One has $supp\ \psi \subset supp\ \varphi$ and $\varphi = x\psi$. Moreover ψ is of class C^∞ for

$$\psi(t) = \int_0^1 \varphi'(tu)du.$$

2) Since $Ker\delta$ is a hyperplane, one has

$$\mathcal{D}(\mathbb{R}) = \mathbb{C}\varphi_0 \oplus Ker\delta, \text{ for every } \varphi_0 \in \mathcal{D}(\mathbb{R}) \text{ such that } \varphi_0(0) = 1$$

and we do have, for any $\varphi \in \mathcal{D}(\mathbb{R})$, $\varphi = \varphi(0)\varphi_0 + \varphi_1$ with $\varphi_1 \in Ker\delta$. But we have seen in **1)** that $\varphi_1 = x\psi_1$ with $\psi_1 \in \mathcal{D}(\mathbb{R})$. It is then clear that $\mathbb{C}\delta \subset KerG$. Conversely let $T \in KerG$ and $\varphi \in \mathcal{D}(\mathbb{R})$. One has $\varphi = \varphi(0)\varphi_0 + \varphi_1$ with $\varphi_1 \in Ker\delta$, and we have, with $\varphi_1 = x\psi_1$

$$\langle T, \varphi \rangle = \langle T, \varphi_0 \rangle \ \langle \delta, \varphi \rangle + \langle xT, \psi_1 \rangle = \langle \langle T, \varphi_0 \rangle \ \delta, \varphi \rangle .$$

Hence $T = \langle T, \varphi_0 \rangle \ \delta$. It remains to show that G is onto. For $S \in \mathcal{D}'(\mathbb{R})$, we look for T such that $xT = S$ i.e., $\langle T, x\varphi \rangle = \langle S, \varphi \rangle$ for every $\varphi \in \mathcal{D}(\mathbb{R})$. But $\varphi = \varphi(0)\varphi_0 + \varphi_1$ with $\varphi_0(0) = 1$ and $\varphi_1(0) = 0$. We must have

$$\langle T, \varphi \rangle = \varphi(0) \langle T, \varphi_0 \rangle + \langle S, \psi_1 \rangle , \text{ where } \varphi_1 = x\psi_1.$$

Replacing T by $T - \langle T, \varphi_0 \rangle \delta$ if necessary, we may suppose $\langle T, \varphi_0 \rangle = 0$. Hence we can put $\langle T, \varphi \rangle = \langle S, \psi_1 \rangle$. One so defines a linear form on $\mathcal{D}(\mathbb{R})$. It is continuous as a composite of continuous maps, taking into account that the hyperplane $Ker\delta$ is closed, and we have $xT = S$. Retain that when considering the equation $xT = S$, one says that it is a question of division over x. $\qquad\qquad\qquad\qquad\qquad\qquad\qquad\qquad\qquad\qquad\qquad$ $\square$

2.7.2. Derivation of distributions

Let $f \in C'(\mathbb{R})$. It is natural to define the derivative $(T_f)'$ of T_f by $T_{f'}$. Integrating by parts, one has

$$\langle (T_f)', \varphi \rangle = - \langle T_f, \varphi' \rangle , \text{ for every } \varphi \in \mathcal{D}(\mathbb{R}).$$

If $f \in \mathcal{L}^1_{loc}(\Omega)$, the right side term is still meaningful. So we are led to the following definition.

Definition 2.4. The partial derivative $\dfrac{\partial T}{\partial x_i}$ of a distribution $T \in \mathcal{D}'(\Omega)$ is given by

$$\left\langle \frac{\partial T}{\partial x_i}, \varphi \right\rangle = - \left\langle T, \frac{\partial \varphi}{\partial x_i} \right\rangle , \ \forall \varphi \in \mathcal{D}(\Omega),$$

and we have

$$\langle D^j T, \varphi \rangle = (-1)^{|j|} \langle T, D^j \varphi \rangle , \text{ for every } \varphi \in \mathcal{D}(\Omega).$$

Remark 2.5. Every distribution is infinitely differentiable and the derivative D^j does not depend of the order of derivations.

Remark 2.6. If T is a distribution of order $\leq m$, then its derivative T' is of order $\leq m+1$.

Example 2.2.

1) Consider the characteristic function of $\mathbb{R}_+$ called Heaviside function and denoted by Y. One has

$$(T_Y)' = \delta \text{ since } \int_0^{+\infty} \varphi'(x)\, dx = -\varphi(0), \text{ for every } \varphi \in \mathcal{D}(\mathbb{R}).$$

For simplicity, we write by abuse of notation, $Y' = \delta$.

2) More generally, let $f \in C(\mathbb{R}\backslash\{a\})$ such that f and f' admit discontinuities of the first kind at the point a. Then

$$(T_f)' = T_{f'} + \left[f(a^+) - f(a^-)\right]\delta_a.$$

Indeed for every $\varphi \in \mathcal{D}(\mathbb{R})$, one has

$$\int_{\mathbb{R}} f(x)\varphi'(x)dx = \int_{-\infty}^{a} f(x)\varphi'(x)dx + \int_{a}^{+\infty} f(x)\varphi'(x)dx$$

$$= -\left[f(a^+) - f(a^-)\right]\varphi(a) - \int_{\mathbb{R}} f'(x)\varphi(x)dx.$$

3) For every $a \in \Omega$, one has

$$\langle D^j \delta_a, \varphi \rangle = (-1)^{|j|} \left(D^j \varphi\right)(a), \quad \varphi \in \mathcal{D}(\mathbb{R}).$$

We are now interested in the range and the kernel of the derivative map.

Proposition 2.11.

1) *The linear map* $d : \varphi \longmapsto \varphi'$, *from* $\mathcal{D}(\mathbb{R})$ *into* $\mathcal{D}(\mathbb{R})$, *is one-to-one and*

$$Im\, d = Ker\, T_1,$$

where T_1 *is the distribution defined by the constant function 1.*

2) *The linear map* $D : T \longmapsto T'$, *from* $\mathcal{D}'(\mathbb{R})$ *into* $\mathcal{D}'(\mathbb{R})$, *is onto and*

$$Ker\, D = \mathbb{C}T_1.$$

Proof.

1) If $d(\varphi) = 0$, then φ is constant. Whence $\varphi = 0$ since it is of compact support. Next it is clear that $Imd \subset KerT_1$. Conversely for $\varphi \in KerT_1$, the function $\psi : x \longmapsto \int_{-\infty}^{x} \varphi(t)dt$ is of class C^{∞} and $\psi' = \varphi$. Moreover ψ is of compact support for if $supp\,\varphi \subset [-a, a]$, with $a > 0$, one has $\psi(x) = 0$ for every $x \notin [-a, a]$.

2) As $KerT_1$ is a hyperplane, one has

$$\mathcal{D}(\mathbb{R}) = \mathbb{C}\varphi_0 \oplus KerT_1, \quad \text{for every } \varphi_0 \in \mathcal{D}(\mathbb{R}),$$

such that

$$\int_{\mathbb{R}} \varphi_0(t)dt = 1.$$

We have, for every $\varphi \in \mathcal{D}(\mathbb{R})$, $\varphi = \left(\int_{\mathbb{R}} \varphi(t)dt \right) \varphi_0 + \varphi_1$ with $\varphi_1 \in KerT_1$. But we have seen in 1) that $\varphi_1 = \psi_1'$ with $\psi_1 \in \mathcal{D}(\mathbb{R})$. It is clear that $\mathbb{C}T_1 \subset KerD$. Conversely for $T \in KerD$ and $\varphi \in \mathcal{D}(\mathbb{R})$, one has

$$\langle T, \varphi \rangle = \left(\int_{\mathbb{R}} \varphi(t)dt \right) \langle T, \varphi_0 \rangle - \langle T', \psi_1 \rangle = \langle \langle T, \varphi_0 \rangle T_1, \varphi \rangle.$$

It remains to show that D is onto. For a given $S \in \mathcal{D}'(\mathbb{R})$, we look for $T \in \mathcal{D}'(\mathbb{R})$ such that $T' = S$, i.e., $\langle T, \varphi' \rangle = -\langle S, \varphi \rangle$ for every $\varphi \in \mathcal{D}(\mathbb{R})$. But $\varphi = \left(\int_{\mathbb{R}} \varphi(t)dt \right) \varphi_0 + \varphi_1$ with $\int_{\mathbb{R}} \varphi_0(t)dt = 1$ and $\int_{\mathbb{R}} \varphi_1(t)dt = 0$. Hence one should have

$$\langle T, \varphi \rangle = \left(\int_{\mathbb{R}} \varphi(t)dt \right) \langle T, \varphi_0 \rangle - \langle S, \psi_1 \rangle \quad \text{where } \psi_1' = \varphi_1.$$

Replacing T by $T - \langle T, \varphi_0 \rangle T_1$ if needed, we may suppose that $\langle T, \varphi_0 \rangle = 0$. Hence we can put $\langle T, \varphi \rangle = -\langle S, \psi_1 \rangle$. A linear form on $\mathcal{D}(\mathbb{R})$ is so defined. It is continuous as a composite of continuous maps, taking into account that the hyperplane $KerT_1$ is closed. So we have $T' = S$. $\square$

As for the product of functions, we have the following result.

Proposition 2.12. *For $T \in (\mathcal{D}^m(\Omega))'$ and $f \in C^m(\Omega)$, one has*

$$\frac{\partial}{\partial x_i}(fT) = \frac{\partial f}{\partial x_i}T + f\frac{\partial T}{\partial x_i}.$$

Remark 2.7. Actually we have the analogue of the Leibniz formula for the product of a function and a distribution.

2.7.3. Linear differential operators on $\mathcal{D}'(\Omega)$

A Partial Differential Equation (PDE in brief) with C^∞-coefficients on an open subset Ω of $\mathbb{R}$ is an expression of the form

$$\sum_{|j|\leq m} a_j D^j T = S, \ S \in D'(\Omega) \text{ given and } a_j \in C^\infty(\Omega).$$

A differential operator with C^∞-coefficients on Ω or a polynomial of derivation with C^∞-coefficients, is an expression of the form

$$P = P(D) = \sum_{|j|\leq m} a_j D^j.$$

The positive integer m is said to be the order of $P(D)$. For every $T \in \mathcal{D}'(\Omega)$, one has $P(D)T = \sum\limits_{|j|\leq m} a_j D^j T$. Hence for every $\varphi \in \mathcal{D}(\Omega)$, we have

$$\langle P(D)T, \varphi \rangle = (-1)^{|j|} \sum_{|j|\leq m} \langle T, D^j(a_j\varphi) \rangle.$$

2.7.4. Significance of the symbol $\langle T, f \rangle$

Recall that if $T \in (\mathcal{D}^m(\Omega))'$ and if $\varphi \in \mathcal{D}^m(\Omega)$, then $\langle T, \varphi \rangle$ designates the value of T at φ. But in certain cases we give a meaning to $\langle T, f \rangle$ without f being with a compact support.

a) **Case where T has compact support.**

We have seen that T extends to $\mathcal{E}^m(\Omega)$ as follows

$$\left\langle \widetilde{T}, f \right\rangle = \langle T, \chi f \rangle, \ \forall f \in \mathcal{E}^m(\Omega),$$

where $\chi \in \mathcal{D}(\Omega)$ with $\chi = 1$ on a neighborhood of the support of T.

b) **Case where supp $f \cap$ supp T is compact.**

Let $f \in \mathcal{E}^m(\Omega)$. We have seen that *supp* $fT \subset$ *supp* $f \cap$ *supp* T. Let $\chi \in \mathcal{D}(\Omega)$ with $\chi = 1$ on a neighborhood of *supp* $f \cap$ *supp* T. Then $\langle T, \chi f \rangle$ has a meaning:

$$\langle T, \chi f \rangle = \langle fT, \chi \rangle = \left\langle \widetilde{fT}, 1 \right\rangle.$$

This leads us to put

$$\langle T, f \rangle = \left\langle \widetilde{fT}, 1 \right\rangle.$$

2.8 Topology in $\mathcal{D}'(\Omega)$

We will endow $\mathcal{D}'(\Omega)$ with the pointwise convergence topology, said also to be weak or vague. It is defined by the family $(p_\varphi)_\varphi$, $\varphi \in \mathcal{D}(\Omega)$, of seminorms where $p_\varphi(T) = |\langle T, \varphi \rangle|$. Hence it is a topology of a Hausdorff *l.c.s.*.

Remark 2.8 (Convergence criterion). *A sequence $(T_n)_n$ in $\mathcal{D}'(\Omega)$ converges to $T \in \mathcal{D}'(\Omega)$ for the simple topology if and only if $p_\varphi(T_n) \xrightarrow[n]{} p_\varphi(T)$ for every $\varphi \in \mathcal{D}(\Omega)$, i.e.,*

$$\forall \varphi \in \mathcal{D}(\Omega), \langle T_n, \varphi \rangle \xrightarrow[n]{} \langle T, \varphi \rangle.$$

We have seen that $\mathcal{L}^1_{loc}(\Omega)$ is algebraically embedded in $\mathcal{D}'(\Omega)$. This injection is continuous. Indeed if a sequence $(f_n)_n$ in $\mathcal{L}^1_{loc}(\Omega)$ converges to $f \in \mathcal{L}^1_{loc}(\Omega)$, then $(T_{f_n})_n$ converges to T_f since, for every $\varphi \in \mathcal{D}(\Omega)$,

$$|\langle T_{f_n}, \varphi \rangle - \langle T_f, \varphi \rangle| \leq \sup_{x \in \Omega} |\varphi(x)| \int_{supp \, \varphi} |f_n(x) - f(x)| \, dx.$$

Hence we have the following result.

Proposition 2.13. *$\mathcal{L}^1_{loc}(\Omega)$ is topologically embedded in $\mathcal{D}'(\Omega)$.*

Examples 2.1 (Examples of convergence). *1) **Gate functions.***
Consider the sequence $(f_n)_{n \geq 1}$ of the so-called Gate functions defined by $f_n = n\chi_{\left[-\frac{1}{2n}, \frac{1}{2n}\right]}$ on $\mathbb{R}$. One has $f_n(x) \xrightarrow[n]{} 0$ for every $x \neq 0$ and $f_n(0) \xrightarrow[n]{} +\infty$. But $T_{f_n} \xrightarrow[n]{} \delta$ since, for every $\varphi \in \mathcal{D}(\mathbb{R})$,

$$\langle T_{f_n}, \varphi \rangle = n \int_{\frac{1}{2n}}^{-\frac{1}{2n}} \varphi(x) dx \xrightarrow[n]{} \varphi(0).$$

Multiplying each f_n by n, one obtains a sequence which still converges almost everywhere to 0 while $(T_{nf_n})_n$ does not converge in $\mathcal{D}'(\mathbb{R})$.

2) Consider on $\mathbb{R}$ the functions f_n, with $n \geq 0$, defined by $f_n(x) = \sin nx$. It is known that $(f_n)_n$ does not have a limit. Integrating by parts, one sees that $\langle T_{f_n}, \varphi \rangle \xrightarrow[n]{} 0$ for every $\varphi \in \mathcal{D}'(\mathbb{R})$.

3) *For every $n \in \mathbb{N}$, consider $f_n(x) = \dfrac{\sin nx}{\pi x}$. The sequence $(f_n)_n$ does
not have a limit while $T_{f_n} \xrightarrow[n]{} \delta$. Indeed let $\varphi \in \mathcal{D}(\mathbb{R})$ and $A > 0$ such
that $supp \, \varphi \subset [-A, A]$. One has $\varphi(x) = \varphi(0) + x\psi(x)$, where ψ is of
class C^∞ such that*

$$\sup_{x \in supp \, \varphi} |\psi(x)| \le c \sup_{x \in supp \, \varphi} |\varphi'(x)|,$$

with a given constant c, and

$$\langle T_{f_n}, \varphi \rangle = \varphi(0) \int_{-A}^{A} \frac{\sin nx}{\pi x} dx + \int_{-A}^{A} \frac{\sin nx}{\pi} \psi(x) dx.$$

Now use

$$\int_{-A}^{A} \frac{\sin nx}{\pi x} dx \xrightarrow[n]{} 1 \ and \ \int_{-A}^{A} \frac{\sin nx}{\pi} \psi(x) dx \xrightarrow[n]{} 0.$$

2.9   Continuity of differential operators

We have the following general result.

Proposition 2.14. *Any differential operator $P(D) = \sum_{|j| \le m} a_j D^j$ with C^∞-
coefficients, on an open set Ω, is a continuous linear map from $\mathcal{D}'(\Omega)$ into
$\mathcal{D}'(\Omega)$.*

Proof.   It suffices to show the continuity of the maps $T \longmapsto fT$, where
$f \in C^\infty(\Omega)$, and $T \longmapsto \dfrac{\partial T}{\partial x_i}$. This follows from the equalities

$$p_\varphi(fT) = p_{f\varphi}(T) \ \text{and} \ p_\varphi\left(\frac{\partial T}{\partial x_i}\right) = p_{\frac{\partial \varphi}{\partial x_i}}(T), \ \varphi \in \mathcal{D}(\Omega). \qquad \square$$

Retain that one has the following inclusions

$$L^1_{loc}(\Omega) \subset [\mathcal{K}(\Omega)]' \subset [\mathcal{D}^k(\Omega)]' \subset \mathcal{D}'(\Omega).$$

Now we have seen (cf. Chapter 1) that $L^1_{loc}(\Omega)$ contains all spaces of functions considered in Chapter 1. So all theses spaces are embedded in $\mathcal{D}'(\Omega)$.

The chapter contains the very basic general properties of distributions.
Most of the results can be found in all books on the theory of distributions. So the difference lies in the presentation. In particular, comments are

provided to clarify the situation and sustain the techniques. We also give more details in the proofs and the presentation of examples. Some examples are usually given as exercises. This is the case, in particular, for the Hadamard finite parts and distributions of infinite order [15]. We have done so, for distribution theory is exactly the frame where their interpretation is appropriate.

It appears that $\mathcal{L}^1_{loc}(\Omega)$ is the bridge between functions and distributions. Indeed, we generally use the regular distributions T_f, with $f \in \mathcal{L}^1_{loc}(\Omega)$, in order to justify the notion to be introduced (support, derivation, product of distributions by a function etc.). This approach is used throughout the book to help the reader fixing notions and results; he could then remember or eventually recapture them. Thus identifying a function f with the associated regular distribution T_f, early, may cause confusion for those not already acquainted with the matter. So we have systematically maintained the notation T_f.

The vague topology on $\mathcal{D}'(\Omega)$, the space of distributions on Ω, is useful. Indeed the convergence of sequences in $\mathcal{D}'(\Omega)$ is actually pointwise, the space $\mathcal{L}^1_{loc}(\Omega)$ is topologically embedded in $\mathcal{D}'(\Omega)$ and moreover any differential operator with C^∞-coefficients is a continuous linear map of $\mathcal{D}'(\Omega)$ into itself.

The properties and the results presented here can not all be found in the same chapter of [14] and [6]; which is for example more or less the case in [3] and [15]. We follow the latter authors considering that this point of view is more enlightening. We still notice that that the division over x (Section 2.7) is always possible for the solution of $xT = 0$ is the straight line $\mathbb{C}\delta$. No detail of this fact is omitted, since it is important in the resolution of equations. The notion of support is related to the order of distributions. Furthermore, distributions with compact support are of particular interest. It is presented here to make it a familiar notion from the outset. For more readings see [3], [4], [5], [7], [8], [14] and [15].

Chapter 3

Tensor Product

This chapter treats the tensor product of distributions. The existence and uniqueness of that product of two arbitrary distributions are shown. We then discuss a very useful particular property, that is the duality formulas by iteration. We also examine the continuity and derivability of the tensor product.

3.1 Tensor product of functions

In the sequel U and V will be two open subsets of $\mathbb{R}^p$ and $\mathbb{R}^q$ respectively.

Definition 3.1. Let f and g be two complex-valued functions defined respectively on U and V. The tensor product of f and g, denoted by $f \otimes g$, is the function defined on $U \times V$ by

$$(f \otimes g)(x, y) = f(x)g(y).$$

If $f \in \mathcal{D}(U)$ and $g \in \mathcal{D}(V)$, then $f \otimes g$ is of class C^∞ since, for every p-tuple α and every q-tuple β of positive integers, one has

$$D^{(\alpha,\beta)}(f \otimes g) = D^\alpha f \otimes D^\beta g,$$

where for $\alpha = (\alpha_1, \ldots, \alpha_p)$ and $\beta = (\beta_1, \ldots, \beta_q)$,

$$D^{(\alpha,\beta)} = \frac{\partial^{|\alpha|+|\beta|}}{\partial x_1^{\alpha_1} \ldots \partial x_p^{\alpha_p} \, \partial y_1^{\beta_1} \ldots \partial y_q^{\beta_q}}.$$

Moreover it has compact support, due to

$$supp \ f \otimes g = supp \ f \times supp \ g.$$

So $f \otimes g \in \mathcal{D}(U \times V)$. We consider the vector subspace of $\mathcal{D}(U \times V)$ generated by the functions $f \otimes g$ with $f \in \mathcal{D}(U)$ and $g \in \mathcal{D}(V)$. We denote it by $\mathcal{D}(U) \otimes \mathcal{D}(V)$.

3.2   Tensor product of distributions

Let $f \in \mathcal{L}^1_{loc}(U)$ and $g \in \mathcal{L}^1_{loc}(V)$. One has $f \otimes g \in \mathcal{L}^1_{loc}(U \times V)$. It is natural to define $T_f \otimes T_g$ by $T_f \otimes T_g = T_{f \otimes g}$. So for $\theta \in \mathcal{D}(U \times V)$ one has

$$\langle T_f \otimes T_g, \theta \rangle = \int_{U \times V} f(x)g(y)\theta(x,y)\, dxdy.$$

If $\theta = \varphi \otimes \psi$ with $\varphi \in \mathcal{D}(U)$ and $\psi \in \mathcal{D}(V)$, then

$$\langle T_f \otimes T_g, \varphi \otimes \psi \rangle = \langle T_f, \varphi \rangle \langle T_g, \psi \rangle.$$

For $\theta \in \mathcal{D}(U \times V)$ arbitrary, we have by Fubini's theorem

$$\langle T_f \otimes T_g, \theta \rangle = \int_U f(x) \left(\int_V g(y)\theta(x,y)dy \right) dx$$
$$= \int_V g(y) \left(\int_U f(x)\theta(x,y)dx \right) dy.$$

Notice that the function

$$x \longmapsto \int_V g(y)\theta(x,y)dy$$

is in $\mathcal{D}(U)$ and that

$$\int_V g(y)\theta(x,y)dy = \langle T_g, \theta(x,.) \rangle, \text{ for every } x \in U.$$

So

$$\langle T_f \otimes T_g, \theta \rangle = \langle T_f, \langle T_g, \theta(.,.) \rangle \rangle.$$

Also

$$\langle T_f \otimes T_g, \theta \rangle = \langle T_g, \langle T_f, \theta(.,.) \rangle \rangle.$$

Another writing is $\langle (T_f)_x, \langle (T_g)_y, \theta(x,y) \rangle \rangle$ instead of $\langle T_f, \langle T_g, \theta(.,.) \rangle \rangle$.

More generally, let $S \in \mathcal{D}'(U)$, $T \in \mathcal{D}'(V)$ and $\theta \in \mathcal{D}(U \times V)$. For any fixed x, the function $y \longmapsto \theta(x,y)$ belongs to $\mathcal{D}(V)$. Hence one can consider $\langle T_y, \theta(x,y) \rangle$. So one obtains the function $x \longmapsto \langle T_y, \theta(x,y) \rangle$. We can apply S to it with the caution that it be in $\mathcal{D}(U)$. This is the matter of the following result.

Theorem 3.1. *Let $T \in \mathcal{D}'(U)$, $\theta \in C^\infty(U \times V)$ and K a compact subset of U such that $\theta(x, y) = 0$ for every $x \notin K$ and every $y \in V$. Then the function*

$$F : y \longmapsto \langle T_x, \theta(x, y) \rangle$$

is of class C^∞ on V. Moreover, for any multi-index $\alpha \in \mathbb{N}^q$, one has

$$D_y^\alpha F = \langle T_x, D_y^\alpha \theta(x, y) \rangle,$$

where D_y^α designates the derivation with respect to y.

Proof. Since the partial derivatives $\dfrac{\partial \theta}{\partial y_i}$ also satisfy the conditions imposed on θ, it suffices to show that the partial derivatives $\dfrac{\partial F}{\partial y_i}$ exist and are given by

$$\frac{\partial F}{\partial y_i}(y) = \left\langle T_x, \frac{\partial \theta}{\partial y_i}(x, y) \right\rangle.$$

We give the proof for the first variable. Let $y_0 \in V$ and $h = (h_1, 0, \ldots, 0)$. One has

$$\frac{F(y_0 + h) - F(y_0)}{h_1} - \left\langle T_x, \frac{\partial \theta}{\partial y_i}(x, y) \right\rangle = \langle T, \psi \rangle$$

with

$$\psi(x) = \frac{\theta(x, y_0 + h) - \theta(x, y_0)}{h_1} - \frac{\partial \theta}{\partial y_1}(x, y_0).$$

Since $\psi \in \mathcal{D}_K(U)$ and $T \in \mathcal{D}'(U)$, there is $c > 0$ and $k \in \mathbb{N}$ such that

$$|\langle T, \psi \rangle| \le c \max_{|\alpha| \le k} \sup_{x \in K} |D^\alpha \psi(x)|.$$

But

$$D^\alpha \psi(x) = \frac{D^\alpha \theta(x, y_0 + h) - D^\alpha \theta(x, y_0)}{h_1} - \frac{\partial}{\partial y_1} D^\alpha \theta(x, y_0).$$

By the mean value theorem, there is $\gamma \in \,]0, 1[$ such that

$$D^\alpha \psi(x) = \frac{\partial}{\partial y_1} D^\alpha \theta(x, y_0 + \gamma h) - \frac{\partial}{\partial y_1} D^\alpha \theta(x, y_0).$$

Put

$$h_\alpha(x, y) = \frac{\partial}{\partial y_1} D^\alpha \theta(x, y).$$

The function h_α is continuous on $U \times V$ and one has $h_\alpha(x, y) = 0$ for $x \notin K$ and every $y \in V$. For α fixed with $|\alpha| \leq k$, let $\varepsilon > 0$ and $x \in U$. There is a neighborhood U_x^α of x in U and a neighborhood V_x^α of y_0 in V such that

$$|h_\alpha(z, y) - h_\alpha(x, y_0)| \leq \frac{\varepsilon}{2c}, \text{ for } z \in U_x^\alpha \text{ and } y \in V_x^\alpha.$$

Then we have

$$|h_\alpha(z, y) - h_\alpha(z, y_0)| \leq \frac{\varepsilon}{c}, \text{ for } z \in U_x^\alpha \text{ and } y \in V_x^\alpha.$$

In view of the compactness of K, there is a neighborhood $V_{y_0}^\alpha$ of y_0 such that

$$|h_\alpha(x, y) - h_\alpha(x, y_0)| \leq \frac{\varepsilon}{c}, \text{ for every } x \in U \text{ and } y \in V_{y_0}^\alpha.$$

Put $V_{y_0} = \bigcap_{|x| \leq k} V_{y_0}^\alpha$. One then has

$$\max_{|\alpha| \leq k} \sup_{x \in U} |h_\alpha(x, y) - h_\alpha(x, y_0)| \leq \frac{\varepsilon}{c}, \text{ for every } y \in V_{y_0}.$$

Hence for h small enough,

$$\max_{|\alpha| \leq k} \sup_{x \in U} |D^\alpha \psi(x)| \leq \frac{\varepsilon}{c}$$

since $D^\alpha \psi(x) = h_\alpha(x, y_0 + \gamma h) - h_\alpha(x, y_0)$. And, for h small enough, we do have,

$$\left| \frac{F(y_0 + h) - F(y_0)}{h_1} - \left\langle T_x, \frac{\partial \theta}{\partial y_1}(x, y_0) \right\rangle \right| \leq \varepsilon. \qquad \square$$

Corollary 3.1. *Let $T \in \mathcal{D}'(U)$ and $\theta \in \mathcal{C}^\infty(U \times V)$.*

(i) If $\theta \in \mathcal{D}_{K \times L}(U \times V)$, then $F \in \mathcal{D}_L(V)$.
(ii) If T has compact support, then $F \in \mathcal{C}^\infty(V)$ and for every multi-index α,

$$D^\alpha F(y) = \langle T, D^\alpha \theta(x, y) \rangle.$$

Proof.

(i) By Theorem 3.1, $F \in \mathcal{C}^\infty(V)$. Moreover it is null outside of L.
(ii) Since T has compact support, one defines F by $F(y) = \langle T, \chi(x)\theta(x, y) \rangle$ where $\chi \in \mathcal{D}(U)$ is such that $\chi = 1$ on a neighborhood of the support of T (cf. 2.7.4 of Chapter 2). It is of class $\mathcal{C}^\infty$ by the previous theorem. Moreover

$$D_y^\alpha F(y) = \langle T, D_y^\alpha \chi(x)\theta(x, y) \rangle = \langle T, \chi(x)D_y^\alpha \theta(x, y) \rangle = \langle T, D_y^\alpha \theta(x, y) \rangle.$$

We will show that there is a distribution and only one on $U \times V$. It is denoted $S \otimes T$, called the tensor product of S by T such that

$$\langle S \otimes T, \varphi \otimes \psi \rangle = \langle S, \varphi \rangle \langle T, \psi \rangle, \quad \varphi \in \mathcal{D}(U), \psi \in \mathcal{D}(V).$$

Uniqueness ensues from the following result. □

Proposition 3.1. *Any distribution T on $U \times V$ vanishing on $\mathcal{D}(U) \otimes \mathcal{D}(V)$ is identically zero.*

Proof. Let $f \in \mathcal{D}(U \times V)$. We have seen (Regularization theorem, Chapter 1) that f is the limit of its regularizations in $\mathcal{D}(U \times V)$. If $(\varphi_j)_j$ and $(\psi_j)_j$ are regularizing sequences in $\mathcal{D}(U)$ and $\mathcal{D}(V)$ respectively, then $\theta_j = \varphi_j \otimes \psi_j$ is a regularizing sequence in $\mathcal{D}(U \times V)$. So $f = \lim_j f * \theta_j$ with $\theta_j \in \mathcal{D}(U) \otimes \mathcal{D}(V)$. Let us show that $\langle T, f * \theta_j \rangle = 0$ for every j. Fix j and let $g = f * \theta_j$. By an approximation theorem of H. Cartan ([2]; see also [15]), $g = \lim_{\varepsilon \to 0} g_\varepsilon$ in $\mathcal{E}(U \times V)$. But here $supp\, f + supp\, \theta_j$ is contained in a compact set which does not depend of ε. Hence one has $g = \lim_{\varepsilon \to 0} g_\varepsilon$ in $\mathcal{D}(U \times V)$ with $g_\varepsilon \in \mathcal{D}(U) \otimes \mathcal{D}(V)$. Then $\langle T, g \rangle = \lim_{\varepsilon \to 0} \langle T, g_\varepsilon \rangle = 0$. □

We now show the existence.

Theorem 3.2. *Let $S \in \mathcal{D}'(U)$ and $T \in \mathcal{D}'(V)$. There is a distribution on $U \times V$, denoted by $S \otimes T$, such that*

$$\langle S \otimes T, \varphi \otimes \psi \rangle = \langle S, \varphi \rangle \langle T, \psi \rangle; \quad \varphi \in \mathcal{D}(U), \psi \in \mathcal{D}(V).$$

Proof. Let $\theta \in \mathcal{D}(U \times V)$. The function defined by $G(y) = \langle S_x, \theta(x, y) \rangle$ is in $\mathcal{D}(V)$. One can consider $\langle T, G \rangle = \langle T_y, \langle S_x, \theta(x, y) \rangle \rangle$. Put $W(\theta) = \langle T_y, \langle S_x, \theta(x, y) \rangle \rangle$. It is a linear form on $\mathcal{D}(U \times V)$. Let us show that it is continuous on $\mathcal{D}(U \times V)$. Let K be a compact subset of U and L a compact subset of V. There are two constants $c > 0$, $c' > 0$ and two integers $k > 0$ and $k' > 0$ such that

$$|\langle S, f \rangle| \le c \max_{|\alpha| \le k} \sup_{x \in U} |D^\alpha f(x)|, \quad \forall f \in \mathcal{D}_K(U)$$

and

$$|\langle T, g \rangle| \le c' \max_{|\beta| \le k'} \sup_{x \in V} |D^\beta g(y)|, \quad \forall g \in D_L(V).$$

Then for every $\theta \in \mathcal{D}_{K \times L}(U \times V)$, one has
$$|W(\theta)| \leq c\, c' \sup_{(x,y)\in K\times L} \max_{|\alpha|\leq k, |\beta|\leq k} \left|D_x^\alpha D_y^\beta \theta(x,y)\right|$$
$$\leq c\, c' \sup_{(x,y)\in K\times L} \max_{|\gamma|\leq k+k'} \left|D^\gamma \theta(x,y)\right|.$$
So W is a distribution on $U \times V$. Finally
$$\langle W, \varphi \otimes \psi \rangle = \langle T_y, \langle S_x, \varphi(x)\psi(y)\rangle\rangle = \langle T_y, \psi(y)\,\langle S_x, \varphi(x)\rangle\rangle = \langle S, \varphi\rangle\,\langle T, \psi\rangle.$$
The previous argument shows that the map $\theta \longmapsto \langle S_x, \langle T_y, \theta(x,y)\rangle\rangle$ is a distribution on $U \times V$ which coincides with $S \otimes T$ on $\mathcal{D}(U) \otimes \mathcal{D}(V)$. $\qquad\square$

Due to the uniqueness of $S \otimes T$, one has the following formulas said to be of duality by iteration. It is the analogue of Fubini theorem for distributions.

Proposition 3.2 (Duality formulas by iteration). *Let $S \in \mathcal{D}'(U)$ and $T \in \mathcal{D}'(V)$. Then for every $\theta \in \mathcal{D}(U \times V)$,*
$$\langle S \otimes T, \theta \rangle = \langle S_x, \langle T_y, \theta(x,y)\rangle\rangle = \langle T_y, \langle S_x, \theta(x,y)\rangle\rangle.$$

We now look at the support of the tensor product of distributions.

Proposition 3.3. *Let $S \in \mathcal{D}'(U)$ and $T \in \mathcal{D}'(V)$. Then*
$$supp\, S \otimes T = (supp\, S) \times (supp\, T).$$

Proof.   First $supp\, S \otimes T \subset (supp\, S) \times (supp\, T)$. This amounts to
$$[(supp\, S)^c \times V] \cup [U \times (supp\, T)^c] \subset (supp S \otimes T)^c.$$
Let $\theta \in \mathcal{D}(U \times V)$. If $supp\, \theta \subset (supp\, S)^c \times V$, then for every $y \in V$, $x \longmapsto \theta(x,y)$ has compact support in $(supp\, S)^c$. Hence $\langle S_x, \theta(x,y)\rangle = 0$ for every $y \in V$. By the duality formulas by iteration, $\langle S \otimes T, \theta\rangle = 0$. The same if $supp\, \theta \subset U \times (supp\, T)^c$. For the converse, let $(a,b) \in (supp\, S) \times (supp\, T)$ and let A and B be two open subsets containing a and b respectively. There are functions $\varphi \in \mathcal{D}(U)$ with $supp\, \varphi \subset A$ and $\psi \in \mathcal{D}(V)$ with $supp\, \psi \subset B$ such that $\langle S, \varphi\rangle \neq 0$ and $\langle T, \psi\rangle \neq 0$. So $\varphi \otimes \psi \in \mathcal{D}(U \times V)$ with $supp\, \varphi \otimes \psi \subset A \times B$ and $\langle S \otimes T, \varphi \otimes \psi\rangle \neq 0$, i.e., $(a,b) \in supp\,(S \otimes T)$. $\qquad\square$

In the case the distributions considered have compact support, the duality formulas are valid even for functions of class $\mathcal{C}^\infty$.

Proposition 3.4. *If $S \in \mathcal{D}'(U)$ and $T \in \mathcal{D}'(V)$ are with compact support, then $S \otimes T$ is with a compact support and for every $\theta \in \mathcal{E}(U \times V)$, one has*
$$\langle S \otimes T, \theta \rangle = \langle S_x, \langle T_y, \theta(x,y)\rangle\rangle = \langle T_y, \langle S_x, \theta(x,y)\rangle\rangle.$$

Proof. The first assertion follows from the previous proposition. For the second, let $\chi_1 \in \mathcal{D}(U)$ with $\chi_1 = 1$ on a neighborhood of *supp* S and let $\chi_2 \in \mathcal{D}(V)$ with $\chi_2 = 1$ on a neighborhood of *supp* T. Then for every $\theta \in \mathcal{E}(U \times V)$, one has

$$\langle S \otimes T, \theta \rangle = \langle S \otimes T, (\chi_1 \otimes \chi_2)\theta \rangle$$

(cf. 2.7.4 of Chapter 2). And by Proposition 3.2,

$$\langle S \otimes T, \theta \rangle = \langle S_x, \chi_1(x) \langle T_y, \chi_2(y)\theta(x,y)\rangle \rangle = \langle T_y, \chi_2(y) \langle S_x, \chi_1(x)\theta(x,y)\rangle \rangle.$$

We finish using 4), 2.7 of Chapter 2 since S and T have compact supports.

$\square$

Remark 3.1. If $U_1, U_2, \cdots, U_m$ are open subsets of $\mathbb{R}^{n_1}$, $\mathbb{R}^{n_2}, \ldots, \mathbb{R}^{n_m}$ respectively, consider the vector subspace $\mathcal{D}(U_1) \otimes \cdots \otimes \mathcal{D}(U_m)$ of $\mathcal{D}(U_1 \times \ldots \times U_m)$ generated by the functions $\varphi_1 \otimes \cdots \otimes \varphi_m$ with $\varphi_i \in \mathcal{D}(U_i)$, $1 \leq i \leq m$.

1) By the same argument as in Proposition 3.1, any distribution on $U_1 \times U_2 \times \cdots \times U_m$ vanishing on $\mathcal{D}(U_1) \otimes \cdots \otimes \mathcal{D}(U_m)$ is identically null.
2) As in Theorem 3.2, one shows that if $T_1 \in \mathcal{D}'(U_1)$, $T_2 \in \mathcal{D}'(U_2), \ldots, T_m \in \mathcal{D}'(U_m)$, then there is a distribution and only one on $U_1 \times U_2 \times \cdots \times U_m$. It is denoted by $T_1 \otimes \cdots \otimes T_m$ and is such that

$$\langle T_1 \otimes \cdots \otimes T_m, \varphi_1 \otimes \cdots \otimes \varphi_m \rangle = \langle T_1, \varphi_1 \rangle \cdots \langle T_m, \varphi_m \rangle.$$

3.3 Properties of tensor product

The tensor product satisfies the following algebraic properties.

Proposition 3.5. *Let* $S \in \mathcal{D}'(U)$ *and* $T \in \mathcal{D}'(V)$.

(i) For every $\theta \in \mathcal{D}(U \times V)$, *one has*

$$\langle S \otimes T, \theta \rangle = \left\langle T \otimes S, \overleftrightarrow{\theta} \right\rangle, \quad \text{where } \overleftrightarrow{\theta}(y,x) = \theta(y,x).$$

Hence the tensor product is not "commutative".

(ii) For every $f \in \mathcal{E}(U)$ *and every* $g \in \mathcal{E}(V)$, *one has*

$$(f \otimes g)(S \otimes T) = (fS) \otimes (gT).$$

(iii) The tensor product is "associative".

Proof.

(i) One has

$$\langle S \otimes T, \theta \rangle = \langle T_y, \langle S_x, \theta(x,y) \rangle \rangle = \left\langle T_y, \left\langle S_x, \overleftrightarrow{\theta}(y,x) \right\rangle \right\rangle = \left\langle S \otimes T, \overleftrightarrow{\theta} \right\rangle.$$

(ii) For every $\varphi \in \mathcal{D}(U)$ and every $\psi \in \mathcal{D}(V)$, one has

$$
\begin{aligned}
\langle (f \otimes g)(S \otimes T), \varphi \otimes \psi \rangle &= \langle S \otimes T, (f \otimes g)(\varphi \otimes \psi) \rangle \\
&= \langle S \otimes T, (f\varphi) \otimes (g\psi) \rangle \\
&= \langle S, f\varphi \rangle \langle T, g\psi \rangle \\
&= \langle fS, \varphi \rangle \langle gT, \psi \rangle \\
&= \langle (fS) \otimes (gT), \varphi \otimes \psi \rangle.
\end{aligned}
$$

Conclude by Proposition 3.1.

(iii) Let $T_1 \in \mathcal{D}'(U)$, $T_2 \in \mathcal{D}'(V)$ and $T_3 \in \mathcal{D}'(W)$. By Remark 3.1, it suffices to show that for $\varphi_1 \in \mathcal{D}(U)$, $\varphi_2 \in \mathcal{D}(V)$ and $\varphi_3 \in \mathcal{D}(W)$,

$$\langle (T_1 \otimes T_2) \otimes T_3, \varphi_1 \otimes \varphi_2 \otimes \varphi_3 \rangle = \langle T_1 \otimes (T_2 \otimes T_3), \varphi_1 \otimes \varphi_2 \otimes \varphi_3 \rangle.$$

Now

$$
\begin{aligned}
\langle (T_1 \otimes T_2) \otimes T_3, \varphi_1 \otimes \varphi_2 \otimes \varphi_3 \rangle &= \langle T_1 \otimes T_2, \varphi_1 \otimes \varphi_2 \rangle \langle T_3, \varphi_3 \rangle \\
&= \langle T_1, \varphi_1 \rangle \langle T_2, \varphi_2 \rangle \langle T_3, \varphi_3 \rangle \\
&= \langle T_1 \otimes (T_2 \otimes T_3), \varphi_1 \otimes \varphi_2 \otimes \varphi_3 \rangle. \qquad \square
\end{aligned}
$$

As for functions, the following result holds.

Proposition 3.6. *Let $S \in \mathcal{D}'(U)$ and $T \in \mathcal{D}'(V)$. Then for $\alpha \in \mathbb{N}^p$ and $\beta \in \mathbb{N}^q$,*

$$D^{(\alpha,\beta)}(S \otimes T) = (D^\alpha S) \otimes (D^\beta T).$$

Proof. For every $\varphi \in \mathcal{D}(U)$ and $\psi \in \mathcal{D}(V)$, one has

$$
\begin{aligned}
\left\langle D^{(\alpha,\beta)}(S \otimes T), \varphi \otimes \psi \right\rangle &= (-1)^{|\alpha|+|\beta|} \left\langle S \otimes T, D^{(\alpha,\beta)}\varphi \otimes \psi \right\rangle \\
&= (-1)^{|\alpha|+|\beta|} \left\langle S \otimes T, D^\alpha\varphi \otimes D^\beta\psi \right\rangle \\
&= (-1)^{|\alpha|}(-1)^{|\beta|} \langle S, D^\alpha\varphi \rangle \langle T, D^\beta\psi \rangle \\
&= \langle D^\alpha S, \varphi \rangle D^\alpha T, \psi \rangle \\
&= \langle (D^\alpha S) \otimes (D^\beta T), \varphi \otimes \psi \rangle. \qquad \square
\end{aligned}
$$

We now examine distributions of finite order.

Proposition 3.7. *If* $S \in \big(\mathcal{D}^h(\Omega)\big)'$ *and* $T \in \big(\mathcal{D}^k(V)\big)'$, *then* $(S \otimes T) \in$ $\big(\mathcal{D}^{h+k}(U \times V)\big)'$ *and for every* $\theta \in \mathcal{D}^{h+k}(U \times V)$, *one has* $\langle S \otimes T, \theta \rangle =$ $\langle S_x, \langle T_y, \theta(x,y) \rangle \rangle$. *In particular, if* S *and* T *are Radon measures, then it is also so for* $S \otimes T$.

Proof. Any compact subset of $U \times V$ is contained in a Cartesian product $K \times L$ where K and L are compact sets in U and V respectively. Let $\theta \in \mathcal{D}_{K \times L}(U \times V)$. We start with $\langle S \otimes T, \theta \rangle = \langle S_x, \langle T_y, \theta(x,y) \rangle \rangle$. Since S is of order less or equal to h, one has

$$|\langle S \otimes T, \theta \rangle| \le c_K \max_{|\alpha| \le h} \sup_{x \in K} |D^\alpha \langle T_y, \theta(x,y) \rangle| \le c_K \max_{|\alpha| \le h} \sup_{x \in K} |\langle T_y, D^\alpha \theta(x,y) \rangle|.$$

Also T is of order less or equal to k, hence

$$|\langle S \otimes T, \theta \rangle| \le c_K\, c_L \sup_{\substack{x \in K\ y \in L \\ |\alpha| \le h |\beta| \le k}} |D^\beta D^\alpha \theta(x,y)|.$$

So

$$|\langle S \otimes T, \theta \rangle| \le c_{K,L} \sup_{\substack{(x,y) \in K \times L \\ |\gamma| \le h+k}} |D^\gamma \theta(x,y)|\,;\ \theta \in \mathcal{D}_{K \times L}(U \times V).$$

For the second assertion, use the duality formulas by iteration (Proposition 3.2) and the regularization theorem (cf. Chapter 1). □

Here is a result on the continuity of tensor product.

Proposition 3.8. *For arbitrary* h *and* k *in* $\overline{\mathbb{N}}$, *the bilinear mapping* $(S,T) \longmapsto S \otimes T$, *from* $\big(\mathcal{D}^h(U)\big)' \times \big(\mathcal{D}^k(V)\big)'$ *into* $\big(\mathcal{D}^{h+k}(U \times V)\big)'$, *is separately continuous, the spaces of distributions* $\big(\mathcal{D}^h(U)\big)'$, $\big(\mathcal{D}^k(V)\big)'$ *and* $\big(\mathcal{D}^{h+k}(U \times V)\big)'$ *being endowed with their vague topologies.*

Proof. Let us show for example that, for every $S \in \big(\mathcal{D}^h(U)\big)'$ fixed, the map $T \longmapsto S \otimes T$ is continuous. For $\theta \in \mathcal{D}^{h+k}(U \times V)$, one has

$$\begin{aligned} p_\theta(S \otimes T) &= |\langle S \otimes T, \theta \rangle| \\ &= |\langle T_y, S_x \langle \theta(x,y) \rangle \rangle| = |\langle T, \psi \rangle| \\ &= p_\psi(T), \text{ where } \psi(y) = \langle S_x, \theta(x,y) \rangle. \end{aligned}$$

And we do have $\psi \in \mathcal{D}^k(V)$. □

The tensor product notion plays an essential role in the definition of convolution product. The first is, in general, considered as a preliminary to the second ([3], [15]). However we preferred to devote a separate chapter for tensor products, as in [14]. So the reader can clearly distinguish the difficulties connected with each one of these products. We are sure that this will be helpful especially for the students. We still mention that tensor products have always their own interest; it appears that the notion is particularly useful in many branches of mathematics.

Concerning the presentation, the method is in effect the rediscovering one. The particular case of regular distributions suggests a formula which may lead to a definition. Then, step by step, we show that this formula extended to arbitrary distributions is exactly the appropriate one; the definition is contained in Theorem 3.2. One also immediately has the duality formulas by iteration which will be basic in the proofs and of frequent use in the sequel. We only show that the tensor product is separately continuous, when the spaces are endowed with the vague topology, that is what we need. More on this can be found in [3],[14] and [15]. Notice that in order to understand the next chapter on the convolution product, this one is an absolute prerequisite.

Chapter 4

Convolution of Distributions

This chapter deals with the convolution product of distributions. This operation is not always possible, which leads to the notion of allowed families for the convolution. Algebraic properties are examined (commutativity, associativity, existence of a neutral element,...). We also look at the derivation and the support. The convolution product of an arbitrary distribution with a regular one is considered too. This allows good density results. Several classical equations (partial differential equations with constant coefficients, finite differences equations with constant coefficients and Volterra equations) are seen as convolution ones in appropriate convolution algebras.

4.1 Regular distributions case

For f, g in $L^1(\mathbb{R}^n)$, the convolution $f * g$ of f and g, given by

$$(f * g)(z) = \int_{\mathbb{R}^n} f(x)g(z-x)dx$$

is in $L^1(\mathbb{R}^n)$. It is natural to define $T_f * T_g$ by T_{f*g}. One has for every $\varphi \in \mathcal{D}(\mathbb{R}^n)$,

$$\langle T_f * T_g, \varphi \rangle = \int_{\mathbb{R}^n} (f * g)(z)\varphi(z)dz$$

$$= \int_{\mathbb{R}^n} \left(\int_{\mathbb{R}^n} f(x)g(z-x)dx \right) \varphi(z)dz.$$

Using Fubini's theorem, one obtains

$$\langle T_f * T_g, \varphi \rangle = \int_{\mathbb{R}^n} \int_{\mathbb{R}^n} f(x)g(z-x)\varphi(z)dxdz$$

$$= \int_{\mathbb{R}^n} \int_{\mathbb{R}^n} f(x)g(y)\varphi(x+y)dxdy.$$

Putting $h(x,y) = f(x)g(y) = (f \otimes g)(x,y)$ and $\varphi^\Delta(x,y) = \varphi(x+y)$, one gets

$$\langle T_f * T_g, \varphi \rangle = \int_{\mathbb{R}^{2n}} h(x,y)\varphi^\Delta(x,y)dxdy.$$

If $f \in L^1_{loc}(\mathbb{R}^n)$ and $g \in L^1_{loc}(\mathbb{R}^n)$, then $h \in L^1_{loc}(\mathbb{R}^{2n})$. But φ^Δ is not necessarily with a compact support. Indeed if $supp\ \varphi = K$, put

$$K^\Delta = (supp\ \varphi)^\Delta = \{(x,y) : x+y \in K\}.$$

If we take $K = [0,1]$, then K^Δ is the strip in $\mathbb{R}^2$ determined by the straight lines the equations of which are $x+y = 0$ and $x+y = 1$. However $\langle T_h, \varphi^\Delta \rangle$ is meaningful when $(supp\ h) \cap (supp\ \varphi^\Delta)$ is compact. But $supp\ h \subset supp\ f \times supp\ g$ and $supp\ (\varphi^\Delta) \subset (supp\ \varphi)^\Delta$. Hence $\langle T_h, \varphi^\Delta \rangle$ has a meaning whenever $(supp\ f \times supp\ g) \cap (supp\ \varphi)^\Delta$ is compact. This is the case in the following situations.

1) $supp\ f$ or $supp\ g$ is compact. Notice that if (A, B) is a couple of subsets of $\mathbb{R}$ one of which is compact and the other closed, and if K is a compact subset of $\mathbb{R}$ then $(A \times B) \cap K^\Delta$ is compact where $K^\Delta = \{(x,y) \in \mathbb{R} \times \mathbb{R} : x+y \in K\}$.
2) $supp\ f$ and $supp\ g$ are subsets of $\mathbb{R}$, both limited on the right (or on the left). Remark that if A and B are included in $]-\infty, a]$ and $K \subset [b,c]$, then

$$(A \times B) \cap K^\Delta \subset ([b-a,a] \times [b-a,a]) \cap K^\Delta.$$

Remark 4.1. If $\langle T_f * T_g, \varphi \rangle$ has a meaning, then
$$\langle T_f * T_g, \varphi \rangle = \langle T_f \otimes T_g, \varphi^\Delta \rangle.$$

4.2 Convolution of distributions

The previous considerations suggest a procedure in order to define the convolution product.

Definition 4.1. Let S and T be two distributions on $\mathbb{R}^n$. The convolution product of S and T, denoted by $S * T$, is a distribution on $\mathbb{R}^n$ (in case of existence) defined by

$$\langle S * T, \varphi \rangle = \langle S \otimes T, \varphi^\Delta \rangle, \ \forall \varphi \in \mathcal{D}(\mathbb{R}^n).$$

As in Section 4.1, one sees that $\langle S * T, \varphi \rangle$ has a meaning whenever $\left[(supp\, S) \times (supp\, T) \right] \cap (supp\, \varphi)^\Delta$ is compact. Now let $A = supp\, S$, $B = supp\, T$ and $K = supp\, \varphi$. One has

$$(A \times B) \cap K^\Delta \subset pr_1 \left[(A \times B) \cap K^\Delta \right] \times pr_2 \left[(A \times B) \cap K^\Delta \right],$$

where pr_1 and pr_2 are the first and the second projections of $\mathbb{R} \times \mathbb{R}$ on $\mathbb{R}$ respectively. Hence it suffices that $pr_1 \left[(A \times B) \cap K^\Delta \right]$ and $pr_2 \left[(A \times B) \cap K^\Delta \right]$ be relatively compact. But

$$pr_1 \left[(A \times B) \cap K^\Delta \right] = A \cap (K - B)$$

and

$$pr_2 \left[(A \times B) \cap K^\Delta \right] = B \cap (K - A).$$

So $\langle S * T, \varphi \rangle$ has a meaning when $supp\, S \cap (supp\, \varphi - supp\, T)$ and $supp\, T \cap (supp\, \varphi - supp\, S)$ are relatively compact. This is the case in particular if $supp\, S$ or $supp\, T$ is compact.

More generally, the convolution product (in case of existence) of k distributions $T_1, \ldots, T_k$ is defined by

$$\langle T_1 * T_2 * \cdots * T_k, \varphi \rangle = \langle T_1 \otimes T_2 \otimes \cdots \otimes T_k, \varphi^\Delta \rangle,$$

where $\varphi^\Delta(u_1, \ldots, u_k) = \varphi(u_1 + \cdots + u_k)$, $u_i \in \mathbb{R}$, $1 \leq i \leq k$. This product is meaningful whenever $\left(\prod_{1 \leq i \leq k} supp\, T_i \right) \cap (supp\, \varphi)^\Delta$ is compact for every $\varphi \in \mathcal{D}(\mathbb{R}^n)$.

The preceding leads to the notion of an allowed family.

Definition 4.2. Let (A_i), $1 \leq i \leq k$, be a family of closed subsets of $\mathbb{R}^n$. We say that it is allowed for the convolution if, for every compact subset K of $\mathbb{R}^n$, the set $\left(\prod_{i=1}^{k} A_i \right) \cap K^\Delta$ is compact in $\mathbb{R}^n$.

Since for any i the i-th projection of $\left(\prod_{i=1}^{k} A_i \right) \cap K^\Delta$ is equal to $A_i \cap \left(K - \sum_{j \neq i} A_j \right)$, we have the following characterization.

Proposition 4.1. *The family* (A_i), $1 \leq i \leq k$, *is allowed for the convolution if and only if for every compact K and $1 \leq i \leq k$, the set* $A_1 \cap \left(K - \sum_{j \neq i} A_j \right)$ *is relatively compact.*

Here is an interesting property (: heredity) of allowed families for the convolution.

Proposition 4.2. *Any subfamily of an allowed family for the convolution is also an allowed one.*

Proof. Let (A_i), $1 \le i \le k$, be an allowed family for the convolution. It suffices to show that for $1 \le i_0 \le k$, the family (A_i), $1 \le i \le k$ and $i \ne i_0$ is also allowed for the convolution. Let $a_{i_0} \in A_{i_0}$ and K an arbitrary compact subset in $\mathbb{R}^n$. Then $L = K + a_{i_0}$ is compact and for every $i \ne i_0$,

$$A_i \cap \left(K - \sum_{l \ne i, i_0} A_l \right) = A_i \cap \left[L - \left(a_{i_0} + \sum_{l \ne i, i_0} A_l \right) \right]$$

$$\subset A_i \cap \left(L - \sum_{l \ne i} A_l \right).$$

$\square$

The following two examples are useful in practice.

(1) a) A family made up of a compact set and an allowed family for the convolution is also allowed for the convolution. Indeed, let $(A_1, \ldots, A_k)$ be an allowed family for the convolution and a compact set A_{k+1}. Then for every $1 \le i \le k$ and any compact K in $\mathbb{R}^n$, one has

$$A_i \cap \left(K - \sum_{1 \le j \le k+1, j \ne i} A_j \right) = A_i \cap \left(K - A_{k+1-} \sum_{1 \le j \le k,\ j \ne i} A_j \right)$$

with $K - A_{k+1}$ compact and $A_{k+1} \cap \left(K - \sum_{1 \le j \le k} A_j \right) \subset A_{k+1}$.

b) Any family made up of a closed set and a family of compact sets is allowed for the convolution. Indeed, by a) it is sufficient to consider a closed set A_1 and a compact set A_2. In that case, (A_1, A_2) is clearly allowed.

(2) Put

$$\Gamma_0 = \{(x_1, \ldots, x_n) \in \mathbb{R}^n : x_n \ge 0\}$$

and for $c > 0$,

$$\Gamma_c = \{(x_1, \ldots, x_n) \in \Gamma_0 : x_n^2 \ge c^2 (x_1^2 + \cdots + x_{n-1}^2)\}, \text{ if } n \ge 2$$
$$\Gamma_c = \Gamma_0, \text{ if } n = 1.$$

One calls Γ_0-set (resp. Γ_c-set) any subset of $\mathbb{R}$ contained in the image of Γ_0 (resp. of Γ_c) by a translation.

Any family of closed subsets of $\mathbb{R}^n$ made up of a Γ_0-set and a family of Γ_c-sets is an allowed family for the convolution. Indeed let $(A_1, \ldots, A_k)$ be a family of Γ_c-sets and A_{k+1} a Γ_0-set. Since a sum of Γ_c-sets is also a Γ_c-set, one has for $1 \leq l \leq k$ and for any compact K of $\mathbb{R}$,

$$A_l \bigcap \left(K - \sum_{1 \leq j \leq k+1,\ j \neq l} A_j \right) \subset A_l \bigcap \left(K - \sum_{1 \leq j \leq k,\ j \neq l} A_j - A_{k+1} \right)$$
$$\subset (\Gamma_c + L) \cap (-\Gamma_0 + L),$$

where L is compact in $\mathbb{R}$. Also

$$A_{k+1} \bigcap \left(K - \sum_{1 \leq j \leq k} A_j \right) \subset (\Gamma_0 + L') \cap (-\Gamma_c + L'),$$

where L' is a compact subset of $\mathbb{R}^n$. The claim amounts to show that for any compact M in $\mathbb{R}^n$, the sets $(\Gamma_0 + M) \cap (-\Gamma_c + M)$ and $(-\Gamma_0 + M) \cap (\Gamma_c + M)$ are compact in $\mathbb{R}$. It is sufficient to proceed for just one of them since $-M$ is also compact. The set $A = (-\Gamma_0 + M) \cap (\Gamma_c + M)$ is closed. It is also bounded for otherwise let $(x_p)_p$ be a non bounded sequence in A. One has

$$x_p = y_p + k_p \in -\Gamma_0 + M \quad \text{and} \quad x_p = z_p + l_p \in \Gamma_c + M$$

with $(k_p)_p$ and $(l_p)_p$ bounded. As $(z_p)_p \subset \Gamma_c$, $z_p = (z_{p,1}, \ldots, z_{p,n})$, we must have a subsequence of $(z_{p,n})_p$ still denoted by $(z_{p,n})_p$ which tends to $+\infty$. Then $(x_{p,n})_p$ tends also to $+\infty$. But this contradicts $x_{p,n} = y_{p,n} + k_{p,n}$ with $y_{p,n} \leq 0$ and $(k_{p,n})_p$ bounded.

Remark 4.2. In the case $n = 1$, one has $\Gamma_c = \Gamma_0 = \mathbb{R}_+$.

Remark 4.3. Let $T_1, T_2, \ldots, T_k$ be distributions on $\mathbb{R}$, the supports of which constitute an allowed family for the convolution. Then for every $\varphi \in \mathcal{D}\left((\mathbb{R}^n)^k \right)$, one has

$$\langle T_1 \otimes T_2 \otimes \ldots \otimes T_k, \varphi^\Delta \rangle = \langle T_1 \otimes T_2 \otimes \ldots \otimes T_k, \chi \varphi^\Delta \rangle$$

for any $\chi \in \mathcal{D}\left((\mathbb{R}^n)^k \right)$ such that $\chi = 1$ on a neighborhood of the compact subset

$$\left(\underset{1 \leq i \leq k}{\Pi}\ supp\ T_i \right) \bigcap (supp\ \varphi)^\Delta.$$

This compact set is contained in the cartesian product

$$\prod_i \left[(supp\ T_i) \bigcap \left(supp\ \varphi - \sum_{j \neq i} supp\ T_j \right) \right]$$

of its projections which are compact. To make calculations easier, choose $\chi = \chi_1 \otimes \cdots \otimes \chi_k$ with $\chi_i \in \mathcal{D}(\mathbb{R}^n)$ and $\chi_i = 1$ on a neighborhood of

$$(supp\ T_i) \bigcap \left(supp\ \varphi - \sum_{j \neq i} supp\ T_j \right).$$

4.3 Properties of the convolution product

We present here some general properties of the convolution product.

For $S, T \in \mathcal{D}'(\mathbb{R}^n)$ and $\varphi \in \mathcal{D}(\mathbb{R}^n)$, one has

$$\langle S * T, \varphi \rangle = \langle S \otimes T, \varphi^\Delta \rangle = \langle S_x \otimes T_y, \varphi(x + y) \rangle.$$

Moreover according to the duality formulas by iteration (Proposition 3.2, Chapter 3), one has

$$\langle S * T, \varphi \rangle = \langle S_x, \langle T_y, \varphi(x, y) \rangle \rangle = \langle T_y, \langle S_x, \varphi(x, y) \rangle \rangle.$$

The convolution product has the following algebraic properties.

Proposition 4.3.

1) *If T_1 and T_2 are distributions the supports of which make up an allowed family for the convolution, then $T_1 * T_2 = T_2 * T_1$.*
2) *For any distribution T on $\mathbb{R}$, one has $T * \delta = \delta * T = T$ where δ is the Dirac measure at the origin.*
3) *If T_1, T_2 and T_3 are distributions the supports of which make up an allowed family for the convolution, then $(T_1 * T_2) * T_3 = T_1 * (T_2 * T_3)$.*

Proof.

1) Follows from the fact that

$$\langle T_1 \otimes T_2, \varphi^\Delta \rangle = \langle T_2 \otimes T_1, \varphi^\Delta \rangle, \quad \forall \varphi \in \mathcal{D}(\mathbb{R}^n).$$

2) As *(supp T, {0})* is an allowed family, $T * \delta$ and $\delta * T$ exist, and we have for every $\varphi \in \mathcal{D}(\mathbb{R}^n)$,

$$\langle \delta * T, \varphi \rangle = \langle \delta, \langle T, \varphi(., .) \rangle \rangle = \langle T, \varphi^\Delta(0, .) \rangle = \langle T, \varphi \rangle.$$

3) Recall that if $\varphi \in \mathcal{D}(\mathbb{R}^n)$ and $x \in \mathbb{R}$, the function $\tau_{-x}\varphi$ is defined by $\tau_{-x}\varphi(t) = \varphi(x + t)$. And we do have

$$\begin{aligned}
\langle T_1 * T_2 * T_3, \varphi \rangle &= \left\langle (T_1)_x \otimes (T_2)_y \otimes (T_3)_z, \varphi(x + y + z) \right\rangle \\
&= \left\langle (T_1)_x \otimes \left((T_2)_y \otimes (T_3)_z \right), \varphi(x + y + z) \right\rangle \\
&= \left\langle (T_1)_x, \left\langle (T_2)_y \otimes (T_3)_z, \tau_{-x}\varphi(y + z) \right\rangle \right\rangle \\
&= \langle (T_1)_x, \langle (T_2 * T_3)_s, \tau_{-x}\varphi(s) \rangle \rangle \\
&= \langle T_1 * (T_2 * T_3), \varphi \rangle.
\end{aligned}$$

In the same way, one shows that

$$\langle T_1 * T_2 * T_3, \varphi \rangle = \langle (T_1 * T_2) * T_3, \varphi \rangle. \qquad \square$$

Concerning the derivation of the convolution product, one has the following.

Proposition 4.4.

1) For any distribution T on $\mathbb{R}^n$, we have

$$D^\alpha \delta * T = T * D^\alpha \delta = D^\alpha T, \ \alpha \in \mathbb{N}^n.$$

2) If T_1 and T_2 are distributions the supports of which make up an allowed family for the convolution, then

$$D^\alpha (T_1 * T_2) = D^\alpha T_1 * T_2 = T_1 * D^\alpha T_2, \ \alpha \in \mathbb{N}^n.$$

Proof.

1) For any $\varphi \in \mathcal{D}(\mathbb{R}^n)$, one has

$$\begin{aligned}
\langle T * D^\alpha \delta, \varphi \rangle &= \left\langle T_x, \left\langle (\delta)_y, (-1)^{|\alpha|} D^\alpha \varphi(x + y) \right\rangle \right\rangle \\
&= \left\langle T, (-1)^{|\alpha|} D^\alpha \varphi \right\rangle = \langle D^\alpha T, \varphi \rangle.
\end{aligned}$$

2) Since $(\{0\}, supp\ T_1, supp\ T_2)$ is an allowed family, $D^\alpha \delta * T_1 * T_2$ exist. Moreover, we have by 1),

$$D^\alpha (T_1 * T_2) = D^\alpha \delta * (T_1 * T_2) = ((D^\alpha \delta) * T_1) * T_2 = D^\alpha T_1 * T_2. \qquad \square$$

We are now interested in the support of the convolution of two distributions.

Proposition 4.5. *Let T_1 and T_2 be two distributions the supports of which make up an allowed family for the convolution. Then $supp\, T_1 + supp\, T_2$ is closed and*

$$supp\, (T_1 * T_2) \subset suppT_1 + suppT_2.$$

Proof. If (A_1, A_2) is an allowed family for the convolution, then $A_1 + A_2$ is closed. Indeed if $c \in \overline{A_1 + A_2}$, there is $(a_p, b_p)_p \subset A_1 \times A_2$ such that $a_p + b_p \xrightarrow{p} c$. The compact set

$$K = \{a_p + b_p;\ p \geq 0\} \cup \{c\}$$

is such that $(a_p, b_p) \in (A_1 \times A_2) \cap K^\Delta$. Now since

$$A_1 \cap (K - A_2) = pr_1\left((A_1 \times A_2) \cap K^\Delta\right),$$

the sequence $(a_p)_p$, in $A_1 \cap (K - A_2)$, is compact. Hence there is a subsequence $(a_{p_k})_k$ of $(a_p)_p$ converging to an element $a \in A_1 \cap (K - A_2)$. Then the sequence $b_{p_k} = (a_{p_k} + b_{p_k}) - a_{p_k}$ converges to $c - a \in \overline{A_2} = A_2$. Hence $c \in a + A_2 \subset A_1 + A_2$. Let us now show that

$$supp\, (T_1 * T_2) \subset suppT_1 + suppT_2.$$

Let $\varphi \in \mathcal{D}(\mathbb{R}^n)$ such that $supp\, \varphi \cap (supp\, T_1 + supp\, T_2) = \emptyset$. Then $supp\, T_1 \cap (supp\, \varphi - supp\, T_2) = \emptyset$. Whence $(supp\, T_1 \times supp\, T_2) \cap (supp\, \varphi)^\Delta = \emptyset$. But since $supp\, T_1 \times supp\, T_2 = supp\, T_1 \otimes T_2$ and $(supp\, \varphi)^\Delta = supp\, (\varphi^\Delta)$, one has $supp\, (T_1 \otimes T_2) \cap supp\, \varphi^\Delta = \emptyset$. Hence $\langle T_1 \otimes T_2, \varphi^\Delta \rangle = 0$, i.e., $\langle T_1 * T_2, \varphi \rangle = 0$. $\qquad\square$

The following particular case is worth mentioning.

Proposition 4.6. *Let $T \in \mathcal{D}'(\mathbb{R}^n)$ and $f \in \mathcal{C}^\infty(\mathbb{R}^n)$ such that the family $(supp\, T, supp\, T_f)$ be allowed for the convolution. Then*

$$T * T_f = T_g,$$

where the function g is of class $\mathcal{C}^\infty$ given by $g(t) = \langle T_x, f(t - x) \rangle$.

Proof. For $\varphi \in \mathcal{D}(\mathbb{R}^n)$, put $\overset{\vee}{\varphi}(u) = \varphi(-u)$. One has

$$\langle T * T_f, \varphi \rangle = \left\langle T_x, \left\langle (T_f)_y, \varphi(x + y) \right\rangle \right\rangle.$$

But

$$\int_{\mathbb{R}^n} f(y)\varphi(x+y)dy = \int_{\mathbb{R}^n} f(t-x)\varphi(t)dt = \int_{\mathbb{R}^n} \overset{\vee}{f}(t+x)\overset{\vee}{\varphi}(t)dt.$$

Hence

$$\begin{aligned}
\langle T * T_f, \varphi \rangle &= \langle T_x, \langle (T_{\overset{\vee}{\varphi}})_t, \overset{\vee}{f}(t+x) \rangle \rangle \\
&= \left\langle (T_{\overset{\vee}{\varphi}})_t, \left\langle T_x, \overset{\vee}{f}(t+x) \right\rangle \right\rangle \\
&= \int_{\mathbb{R}^n} \overset{\vee}{\varphi}(t) \left\langle T_x, \overset{\vee}{f}(t+x) \right\rangle dt \\
&= \int_{\mathbb{R}^n} g(t)\varphi(t)dt = \langle T_g, \varphi \rangle.
\end{aligned}$$

$\square$

Definition 4.3. The distribution $T * T_f$ is called the convolution of T and f. It is also denoted by $T * f$. It is then said to be the regularization of T by f.

As a consequence, we have the following density results.

Proposition 4.7. *Any distribution T is a limit in $\mathcal{D}'(\mathbb{R}^n)$ of a sequence of functions of class $\mathcal{C}^\infty$; we say that $\mathcal{E}(\mathbb{R}^n)$ is dense in $\mathcal{D}'(\mathbb{R}^n)$.*

Proof. Remark that for $\varphi \in \mathcal{D}(\mathbb{R}^n)$, one has $\left(T * \overset{\vee}{\varphi}\right)(0) = \langle T, \varphi \rangle$. Let $(\theta_j)_j$ be a regularizing sequence. Then

$$\langle T * \theta_j, \varphi \rangle = \left[(T * \theta_j) * \overset{\vee}{\varphi} \right](0) = \left[\theta_j * \left(T * \overset{\vee}{\varphi}\right) \right](0).$$

But $\theta_j * \left(T * \overset{\vee}{\varphi}\right) \underset{j}{\longrightarrow} T * \overset{\vee}{\varphi}$. Hence $\langle T * \theta_j, \varphi \rangle \underset{j}{\longrightarrow} \left(T * \overset{\vee}{\varphi}\right)(0) = \langle T, \varphi \rangle.$ $\square$

As we have seen that $\mathcal{D}(\mathbb{R}^n)$ is dense in $\mathcal{E}(\mathbb{R}^n)$, one also has the following result.

Proposition 4.8. *The space $\mathcal{D}(\mathbb{R}^n)$ is dense in $\mathcal{D}'(\mathbb{R}^n)$.*

4.4   Convolution algebras

One can not always define the convolution product of two arbitrary distributions. Take for example $T_1 = T_Y$ and $T_2 = T_1$, where Y is the Heaviside function and $\mathbf{1}$ the constant function equal to 1. We have seen that in $\mathcal{E}'(\mathbb{R}^n)$ the convolution product is always defined and has good properties. This is also the case for the space $\mathcal{D}'^{+}(\mathbb{R})$ of distributions on $\mathbb{R}$ with support in $\mathbb{R}^+$. Actually $\mathcal{E}'(\mathbb{R}^n)$ and $\mathcal{D}'^{+}(\mathbb{R})$ are endowed with an algebra structure.

Definition 4.4. We say that a subset $\mathcal{A} \subset \mathcal{D}'(\mathbb{R}^n)$ is a convolution algebra (said to be unital if $\delta \in \mathcal{A}$) when

(i) $\mathcal{A}$ is a vector subspace of $\mathcal{D}'(R^n)$,
(ii) the convolution product is defined in $\mathcal{A}$,
(iii) the convolution product in $\mathcal{A}$ is commutative and associative.

Example 4.1.

1) The space $\mathcal{E}'(\mathbb{R}^n)$ is a unital convolution algebra.
2) The space of distributions with Γ_c-set supports (supports limited on the left if $n = 1$) is a unital convolution algebra.
3) The spaces $\{T_f, f \in \mathcal{D}(\mathbb{R}^n)\}, \{T_f, f \in \mathcal{K}(\mathbb{R}^n)\}$ and $\{T_f, \ f \in L^1(\mathbb{R}^n)\}$ are non unital convolution algebras.

Moreover in the convolution algebras $\mathcal{E}'(\mathbb{R}^n)$ and $\mathcal{D}'^{+}(\mathbb{R})$ the convolution product satisfies the following interesting property.

Proposition 4.9. *In the convolution algebras $\mathcal{E}'(\mathbb{R}^n)$ and $\mathcal{D}'_{+}(\mathbb{R})$, the convolution product is separately continuous, i.e., for any fixed S, the map $T \longmapsto T * S$ is continuous.*

Proof.   Recall first that for every $\varphi \in \mathcal{D}(\mathbb{R}^n)$, one has
$$\langle S * T, \varphi \rangle = \langle T_x, \langle S_y, \varphi(x+y) \rangle \rangle = \langle T_x, \psi(x) \rangle$$
with $\psi(x) = \langle S_y, \varphi(x+y) \rangle$. The function ψ is of class $\mathcal{C}^\infty$. Moreover, one has $supp\ \psi \subset supp\ \varphi - supp\ S$. In the case of $\mathcal{E}'(\mathbb{R}^n)$, one has $\psi \in \mathcal{D}(\mathbb{R}^n)$ and $p_\varphi(S * T) = p_\psi(T)$. In the case of $\mathcal{D}'^{+}(\mathbb{R})$, let $a < 0 < b$ be such that $supp\ \varphi \subset [a, b]$ and let U be an open set containing $[a, b]$. There is $\chi \in \mathcal{D}(\mathbb{R})$ such that $\chi = 1$ on $[a, b]$ with $supp\ \chi \subset U$. Since $supp\ T \cap supp\ (\psi - \chi\psi) = \emptyset$, one has $\langle T, \psi \rangle = \langle T, \chi\psi \rangle$, $p_\varphi(S * T) = p_{\chi\psi}(T)$.   □

4.5 Convolution equations

If we consider a convolution algebra $\mathcal{A}$ then for A, B and X in $\mathcal{A}$, the equation $A * X = B$ has a meaning. But for $A \in \mathcal{A}$, $A * X$ can have a meaning even if $X \notin \mathcal{A}$; in this case $A * X$ does not necessarily belong to $\mathcal{A}$. Hence one may look for solutions in a subset of $\mathcal{D}'(\mathbb{R}^n)$ containing $\mathcal{A}$. This leads us to the notion of a module.

Definition 4.5. Let $\mathcal{A}$ be a unital convolution algebra and let $\mathcal{X}$ be a vector subspace of $\mathcal{D}'(\mathbb{R}^n)$ containing $\mathcal{A}$. We say that $\mathcal{X}$ is a convolution module over $\mathcal{A}$ if

(i) For every $A \in \mathcal{A}$ and every $X \in \mathcal{X}$, both $A * X$ and $X * A$ exist, are in $\mathcal{X}$ and $A * X = X * A$.
(ii) For any A, $B \in \mathcal{A}$ and any $X \in \mathcal{X}$, one has $(A * B) * X = A * (B * X)$.
(iii) For any A, $B \in \mathcal{A}$ and any $X \in \mathcal{X}$, one has $(A + B) * X = A * X + B * X$.
(iv) For every $A \in \mathcal{A}$ and every $X, Y \in \mathcal{X}$, one has $A * (X + Y) = A * X + A * Y$.

Example 4.2.

1) Any unital convolution algebra is a convolution module over itself.
2) The space $\mathcal{D}'(\mathbb{R}^n)$ is a convolution module over $\mathcal{E}'(\mathbb{R}^n)$.
3) The space of distributions the supports of which are Γ_0-sets is a convolution module over the convolution algebra of distributions the supports of which are Γ_c-sets.

Definition 4.6. Let $\mathcal{X}$ be a convolution module over $\mathcal{A}$. A convolution equation in $\mathcal{X}$ is any expression of the form

$$A * X = B; \quad A \in \mathcal{A}, \ B \in \mathcal{X} \text{ and } X \in \mathcal{X}.$$

We say that A is the coefficient (or the convolution kernel) of the equation, that B is its right-hand side and X the unknown.

Example 4.3.

1) A linear partial differential equation with constant coefficients (cf. 2.7.3 of Chapter 2)

$$\sum_{|j| \leq m} \alpha_j D^j X = B \qquad (1)$$

can be viewed as a convolution equation in $\mathcal{D}'(\mathbb{R}^n)$. Indeed it can be written

$$\left(\sum_{|j|\leq m} \alpha_j D^j \delta\right) * X = B$$

the coefficients α_j are C^∞ functions with compact supports.

2) We consider an equation with finite differences and constant coefficients

$$\sum_{i=1}^{N} \alpha_{a_i} \tau_{a_i} X = B \tag{2}$$

where $\langle \tau_a X, \varphi \rangle = \langle X, \tau_{-a}\varphi \rangle$, $a \in R^n$ and $(\tau_b \varphi)(x) = \varphi(x - b)$, $b \in R^n$. It can be considered as a convolution equation in $D'(R^n)$. Indeed it can be written

$$\left(\sum_{i=1}^{N} \alpha_{a_i} \delta_{a_i}\right) * X = B.$$

3) **Volterra equation of the first kind**

For g and K locally integrable functions with support in $\mathbb{R}_+$, the equation with f as unknown

$$f(x) + \int_0^x K(x-t)f(t)dt = g(x) \tag{3}$$

can be viewed as a convolution equation in $\mathcal{D}'^+(\mathbb{R})$. Indeed it can be written

$$(\delta + T_K) * f = g$$

for

$$(\delta * f)(x) = f(x) \text{ and } (T_K * f)(x) = (K * f)(x) = \int_0^x K(x-t)f(t)dt.$$

4) **Volterra equation of the second kind**

For g and K be locally integrable functions with compact support in $\mathbb{R}_+$, the equation with f as unknown

$$\int_0^x K(x-t)f(t)dt = g(x) \tag{4}$$

can be viewed as a convolution equation in $\mathcal{D}'^+(\mathbb{R})$.

The following general results will be very useful in the sequel.

Proposition 4.10. *Let $A * X = B$ be a convolution equation in $\mathcal{X}$. If the vector space $\mathcal{X}$ is endowed with a separated t.v.s. topology such that the map $X \longmapsto A * X$ is continuous, then the set of solutions of the equation is closed in $\mathcal{X}$.*

Proof. Ensues from the continuity of the map $X \longmapsto A * X - B$. □

Remark 4.4. If $A \in \mathcal{E}'(\mathbb{R}^n)$, then the set of solutions of the homogeneous equation $A*X = 0$ in $\mathcal{D}'(\mathbb{R}^n)$ is a closed vector subspace of $\mathcal{D}'(\mathbb{R}^n)$. Indeed, the same proof as for 1) in Proposition 4.3 shows that $X \longmapsto A * X$, from $\mathcal{D}'(\mathbb{R}^n)$ into $\mathcal{D}'(\mathbb{R}^n)$, is continuous. It is the same for the equation $A*X = 0$ in $D'_+(\mathbb{R})$ if $A \in D'_+(\mathbb{R})$.

Proposition 4.11. *If $A \in \mathcal{E}'(\mathbb{R}^n)$, then the set of solutions of $A*X = 0$ in $\mathcal{D}'(\mathbb{R}^n)$ is the closure in $\mathcal{D}'(\mathbb{R}^n)$ of the set of solutions which are functions of class $\mathcal{C}^\infty$.*

Proof. If $T \in \mathcal{D}'(\mathbb{R}^n)$ is a solution, then any regularization $T_j = T * \theta_j$ is also a solution. We then apply Proposition 4.7. □

An interesting particular case is the following.

Proposition 4.12. *In $\mathcal{D}'(\mathbb{R}^n)$, the solutions of a homogeneous differential equation with constant coefficients are (all of them) regular distributions associated with functions of class $\mathcal{C}^\infty$; they are said then, by abuse of language, that coincide with the usual ones.*

Proof. The space of solutions of class $\mathcal{C}^\infty$ is a vector space of finite dimension. Hence it is closed for the topology induced by that one of $\mathcal{D}'(\mathbb{R}^n)$. One concludes using Proposition 4.11. □

In a module $\mathcal{X}$ over an algebra $\mathcal{A}$, the equation $A * X = \delta$ may admit a solution X which does not belong to $\mathcal{A}$. For example one has $\delta' * Y = \delta$ in $\mathcal{D}'(\mathbb{R}^n)$ over $\mathcal{E}'(\mathbb{R}^n)$.

Definition 4.7. A fundamental solution of $A * X = B$, is any solution X of the equation $A * X = \delta$.

Remark 4.5.

1) If $f \in \mathcal{D}(\mathbb{R}^n)$, then the equation $f * X = \delta$ admits no solution. Indeed $f * X \in \mathcal{C}^\infty(\mathbb{R})$ and we have seen that δ is not regular. So $A * X = B$ does not always admit a fundamental solution.
2) A solution of the equation $A * X = B$ is not unique in general. Indeed the sum of a fundamental solution and a solution of the homogeneous equation is yet a fundamental solution.

Proposition 4.13. *If $A * X = B$ admits fundamental solutions, then any solution is of the form $E * B + Z$, where E is a fundamental solution and Z a solution of the homogeneous equation.*

Proof. Notice that $E * B$ is a solution of $A * X = B$ and that if X is any solution, then $X - E * B = Z$ is a solution of $A * X = 0$. □

Concerning uniqueness, one has the following.

Proposition 4.14. *If $A * X = B$ admits a fundamental solution belonging to the algebra $\mathcal{A}$, then*

(i) The fundamental solution E is unique in $\mathcal{X}$.
(ii) The equation admits a solution and only one, that is $E * B$.

Proof.

(i) If $A * F = \delta$ with $F \in \mathcal{X}$, one has
$$E = E * \delta = E * (A * F) = (E * A) * F = \delta * F = F.$$
(ii) By Proposition 4.13, this amounts to the fact that the homogeneous equation admits only the trivial solution. Now if $A * X = 0$, then
$$X = \delta * X = (E * A) * X = E * (A * X) = 0.$$
□

Remark 4.6. The equation $A * X = B$ may admit an infinity of fundamental solutions. Take for example $\delta' * X = \delta$ in $\mathcal{D}'(\mathbb{R})$.

Now we consider the case of an ordinary differential equation, i.e., an equation in $\mathcal{D}'^+(\mathbb{R})$ of the form
$$\left(\delta^{(m)} + \alpha_{m-1}\delta^{(m-1)} + ... + \alpha_1\delta + \alpha_0\right) * X = B,$$
with complex coefficients. The study is reduced to the search of fundamental solutions, i.e., to solve in $\mathcal{D}'^+(\mathbb{R})$ the equation
$$E^{(m)} + \alpha_{m-1}E^{(m-1)} + ... + \alpha_1 E + \alpha_0 = \delta.$$

We try to find solutions, if any, of the form T_h with $h \in L^1_{loc}(\mathbb{R})$ and *supp* $h \subset \mathbb{R}_+$. Good candidates are $h = Yf$ where Y is the Heaviside function and $f \in \mathcal{C}^\infty(\mathbb{R})$. One then has

$$(Yf)' = Yf' + f(0)\delta$$
$$(Yf)'' = Yf'' + f'(0)\delta + f(0)\delta'$$

$$\vdots \quad \vdots \quad \vdots$$

$$(Yf)^{(m)} = Yf^{(m)} + f^{(m-1)}(0)\delta + f^{(m-2)}(0)\delta' + \cdots + f(0)\delta^{(m-1)}.$$

Hence

$$\delta = \left[\alpha_0 f + \alpha_1 f' + \cdots + \alpha_{m-1}f^{(m-1)} + f^{(m)}\right]Y$$
$$+ \left[\alpha_1 f(0) + \alpha_2 f'(0) + \cdots + \alpha_{m-1}f^{(m-2)}(0) + f^{(m-1)}(0)\right]\delta$$
$$+ \left[\alpha_2 f(0) + \alpha_3 f'(0) + \cdots + \alpha_{m-1}f^{(m-3)}(0) + f^{(m-2)}(0)\right]\delta'$$
$$+$$
$$\vdots$$
$$+ \left[\alpha_{m-1}f(0) + f'(0)\right]\delta^{(m-2)}$$
$$+ f(0)\delta^{(m-1)}.$$

Since for any nonzero function $g \in \mathcal{C}^\infty(\mathbb{R})$, the family $(T_{Yg}, \delta, \delta', \ldots, \delta^{(m-1)})$ is free, it follows that

$$\alpha_0 f + \alpha_1 f' + \cdots + \alpha_m f^{(m)} = 0,$$

$$f(0) = f'(0) = \cdots = f^{(m-2)}(0) = 0 \text{ and } f^{(m-1)}(0) = 1.$$

So we have the following result.

Proposition 4.15. *An ordinary differential equation with constant coefficients*

$$E^{(m)} + \alpha_{m-1}E^{(m-1)} + \ldots + \alpha_1 E + \alpha_0 = \delta$$

admits in $\mathcal{D}'^+(\mathbb{R})$ a solution and only one. It is of the form T_{Yf} where f is a function of class $\mathcal{C}^\infty$, a solution of the homogeneous equation, such that

$$f(0) = f'(0) = \cdots = f^{(m-2)}(0) = 0 \text{ and } f^{(m-1)}(0) = 1.$$

Remark 4.7. Propositions 4.14 and 4.15 show that in $\mathcal{D}'^+(\mathbb{R})$ the equation $P(D)A = B$ admits a solution and only one, that is $T = E * B$. Moreover, the homogeneous equation admits only the trivial solution.

In the case B is a regular distribution, one has the following result.

Proposition 4.16. *Consider in $\mathcal{D}'^+(\mathbb{R})$ the equation*

$$P(D)A = T_v \text{ with } v \in L^1_{loc}(\mathbb{R}) \text{ and supp } v \subset \mathbb{R}_+.$$

(i) The solution of this equation is a regular distribution given by a function of class $\mathcal{C}^{m-1}$ on $\mathbb{R}$.

(ii) If v is of class $\mathcal{C}^k$, then the solution is given by a function of class $\mathcal{C}^{m+k}$ on $\mathbb{R}$.

Proof.   One knows that the solution is $T_{Yf} * T_v$. It is equal to $T_{Yf} * v$.

(i) First $Yf * v$ is continuous. For this show that for every compact set K in $\mathbb{R}$, there is a function g_K continuous on $\mathbb{R}$ such that $g_K = Yf * v$ on K. For $t \in K$, one checks that

$$(Yf * v)(t) = Yf * v\chi_{K-suppYf}(t)$$

and that $v\chi_{K-suppYf}$ has compact support. Since $Yf \in L^\infty_{loc}(\mathbb{R})$ and $v\chi_{K-suppYf} \in L^1(\mathbb{R})$, the function $Yf * v\chi_{K-suppYf}$ is continuous on $\mathbb{R}$. For $j \leq m-1$, one has $D^j T_{Yf} = T_{Yf^{(j)}}$. Hence $D^j(T_{Yf*v}) = T_{Yf^{(j)}*v}$. As $Yf^{(j)} \in L^\infty_{loc}(\mathbb{R})$ the previous argument shows that $f^{(j)} * v$ is continuous. One concludes using the lemma of Du Bois-Reymond (cf. [15]).

(ii) Notice first that the equation $P(D)A = T_v$ gives $D^m A = T_g$ where

$$g = \frac{1}{\alpha_m}\left[v - \left(\alpha_{m-1}Yf^{(m-1)} * v + \cdots + \alpha_o Yf * v\right)\right]$$

for $A = T_{Yf*v}$. So if v is continuous, then g is also continuous since $Yf^{(j)} * v$ is continuous for $0 \leq j \leq m-1$. Then, by **(i)** and the lemma of Du Bois-Reymond (cf. [15]), A is given by a function of class $\mathcal{C}^m$. If v is of class $\mathcal{C}^k$ one has for $j \leq k$, $P(D)(D^j A) = T_{D^j v}$ since $D^j(T_v) = T_{D^j v}$. But $D^k v$ being continuous $D^k A$ is given by a function of class $\mathcal{C}^m$. Using again the lemma of Du Bois-Reymond (*ibid.*) one sees that A is given by a function of class $\mathcal{C}^{m+k}$. $\square$

We will apply the above to solve the Cauchy problem, i.e., to look for the solution of the equation (in the framework of distributions)

$$P(D)u = v \text{ on }]0, +\infty[\tag{1}$$

where

$$P(D) = \alpha_m \frac{d^m}{dx^m} + \cdots + \alpha_1 \frac{d}{dx} + \alpha_0$$

and $v \in \mathcal{L}^1_{loc}(\mathbb{R})$ is a given function and u the unknown with the initial conditions

$$u^{(k)}(0) = a_k$$

for $0 \le k \le m - 1$.

Proposition 4.17. *The Cauchy problem admits a solution T_u and only one, u being a function of class $\mathcal{C}^{m-1}$ on $\mathbb{R}^+$ given by*

$$u(t) = \int_0^t f(t - s)v(s)ds + \sum_{j=0}^m b_j f^{(j)}(t)$$

where

$$b_j = \alpha_{j+1}a_o + \alpha_{j+2}a_1 + \cdots + \alpha_m a_{m-(j+1)}$$

and f the function in Proposition 4.15.

Proof. Let u be a solution. Consider the function w defined on $\mathbb{R}$ by $w(t) = u(t)$ if $t \ge 0$ and $w(t) = 0$ if $t < 0$. Then

$$w = Yu = u \text{ on } \mathbb{R}_+.$$

Differentiating in the context of distributions, one has

$$Y_u = u$$

$$\vdots$$

$$(Y_u)^{(j)} = Yu^{(j)} + u^{(j-1)}(0)\delta + \cdots + u'(0)\delta^{(j-2)} + u(0)\delta^{(j-1)}$$

$$\vdots$$

$$(Y_u)^{(m)} = Yu^{(m)} + u^{(m-1)}(0)\,\delta + \cdots + u'(0)\,\delta^{(m-2)} + u(0)\,\delta^{(m-1)}$$

hence we do have

$$P(D)(T_u) = Yv + \sum_{j=0}^{m-1} b_j \delta^{(j)}. \tag{2}$$

By Proposition 4.13, one has

$$T_u = T_{Yf} * \left(T_{Yv} + \sum_{j=0}^{m-1} b_j \delta^{(j)} \right) = T_{Yf} * T_{Yv} + \sum_{j=0}^{m-1} b_j \left(T_{Yf} \right)^{(j)}.$$

According to the previous proposition, $T_{Yf} * T_{Yv}$ is a function of class $\mathcal{C}^{m-1}$ on $\mathbb{R}$. Moreover, $T_{Yf} * T_{Yv} = T_{Yf} * Yv$. And for $t \geq 0$

$$(Yf * Yv)(t) = \int_o^t f(t-s)v(s)ds.$$

On the other hand, for $0 \leq j \leq m-1$, the restriction of $(T_{Yf})^{(j)}$ to $]0, +\infty[$ is equal to $T_{f^{(j)}}$. So on the open interval $]0, +\infty[$ one has $T_u = T_g$ with

$$g(t) = \int_o^t f(t-s)v(s)ds + \sum_{j=0}^{m-1} b_j f^{(j)}(t).$$

This function is in fact of class $\mathcal{C}^{m-1}$ on $\mathbb{R}_+$. The restriction of **(2)** to $]0, +\infty[$ is written $P(D)T_u = T_v$. So T_g is a solution of **(1)**. Finally u satisfies the initial conditions. Indeed **(2)** is also written

$$P(D)T_u = Yv + \sum_{j=0}^{m-1} b_j^\sharp \delta^{(j)}$$

where

$$b_j^\sharp = \alpha_{j+1}u(0) + \cdots + \alpha_m u^{[m-(j+1)]}(0).$$

Whence $b_j = b_j^\sharp$ for $0 \leq j \leq m-1$, due to the fact that $\left(\delta^{(j)} \right)_{0 \leq j \leq m-1}$ is free. It folows then that $a_j = u^{(j)}(0)$ for $0 \leq j \leq m-1$. $\square$

Remark 4.8. If v is continuous, then g is of class $\mathcal{C}^m$ by **(ii)** of Proposition 4.16. The equality $P(D)T_g = T_v$ is then written $T_{P(D)g} = T_v$. Hence one has $P(D)g = v$ for the two functions are continuous. So g is a solution of **(1)** in the usual sense.

Example 4.4.

1) Take

$$P(D) = \frac{d}{dx} + \lambda, \quad \lambda \in \mathcal{C}.$$

By Proposition 4.15, the equation $P(D)A = \delta$ has as a solution T_{Yf} in $\mathcal{D}'^+(\mathbb{R})$ where f is the solution of the equation $f' + \lambda f = 0$ with

$f(0) = 1$. Hence the solution is $T_{Ye^{-\lambda x}}$ for $f(x) = e^{-\lambda x}$. The equality $P(D)T_{Ye^{-\lambda x}} = \delta$ is also written $(\delta' + \lambda\delta) * T_{Ye^{-\lambda x}} = \delta$. So $(\delta' + \lambda\delta)$ is invertible in $\mathcal{D}'^+(\mathbb{R})$; and one has

$$(\delta' + \lambda\delta)^{-1} = T_{Ye^{-\lambda x}}.$$

2) Take

$$P(D) = \frac{d^2}{dx^2} + \omega^2, \quad \omega \in \mathbb{R}.$$

In $\mathcal{D}'^+(\mathbb{R})$, the solution of $P(D)A = \delta$ is $A = T_{Y\frac{\sin \omega x}{\omega}}$ for the solution of $f'' + \omega^2 f = 0$, with the initial conditions $f(0) = 0$ and $f'(0) = 1$, is $f(x) = \frac{\sin \omega x}{\omega}$. Here one has

$$\left(\delta'' + \omega^2\delta\right)^{-1} = T_{Y\frac{\sin \omega x}{\omega}}.$$

3) Take

$$P(D) = \alpha_m \frac{d^m}{dx^m} + \cdots + \alpha_1 \frac{d}{dx} + \alpha_0; \quad \alpha_i \in \mathcal{C}, \ i = 0, \ldots, m.$$

The equation $P(D)A = \delta$ is written $\left(\sum_{i=0}^m \alpha_i\delta^{(i)}\right) * A = \delta$. But

$$\sum_{i=0}^m \alpha_i\delta^{(i)} = \alpha_m(\delta' - z_1\delta) * \cdots * (\delta' - z_m\delta)$$

where $z_1, \ldots, z_m$ are the roots of the polynomial $\alpha_m z^m + \cdots + \alpha_1 z + \alpha_0$. The solution of the equation $P(D)A = \delta$ in $\mathcal{D}'^+(\mathbb{R})$ is

$$\frac{1}{\alpha_m}\left[(\delta' - z_1\delta) * \cdots * (\delta' - z_m\delta)\right]^{-1}$$

$$= \frac{1}{\alpha_m}(\delta' - z_1\delta)^{-1} * \cdots * (\delta' - z_m\delta)^{-1}$$

$$= \frac{1}{\alpha_m}T_{Ye^{z_1 x}} * \cdots * T_{Ye^{z_m x}}$$

$$= T_{\frac{1}{\alpha_m}Ye^{az_1 x} * \ldots * Ye^{z_m x}}.$$

In particular if $z_1 = \cdots = z_m = \lambda$, then the solution becomes

$$T_{\frac{1}{\alpha_m}Y\frac{x^{m-1}}{(m-1)!}e^{\lambda x}}$$

for

$$Ye^{\lambda x} * \cdots * Ye^{\lambda x} = Y\frac{x^{m-1}}{(m-1)!}e^{\lambda x}.$$

The convolution product is a very important operation in the theory of distributions with several applications. It is known that the ordinary product of two non zero distributions S and T is not defined. Actually ST is linear if, and only if, S or T is zero.

A way to introduce the convolution product can be deduced from calculations on regular distributions, by also involving tensor product. However, one cannot always define the convolution product of two arbitrary distributions (cf. Section 4.4). This leads to the fundamental notion of allowed families. Thus we can obtain the density of $\mathcal{D}(\mathbb{R}^n)$ in $\mathcal{D}'(\mathbb{R}^n)$, using the regularizations $T * f$ of a distribution T by functions f. The strength of this operation is that $T * f$, $f \in \mathcal{D}(\mathbb{R}^n)$, is a regular distribution T_g, with $g \in \mathcal{D}(\mathbb{R}^n)$, for any distribution T. We thus have the same phenomenon as encountered in the regularizing method of functions presented in Chapter 1.

From properties of allowed families (cf. Examples (1) and (2), Section 4.2), one notices that the subspaces $\mathcal{E}'(\mathbb{R}^n)$ and $\mathcal{D}'_+(\mathbb{R}^n)$ of $\mathcal{D}'(\mathbb{R}^n)$ are stable by the convolution product being therefore convolution algebras. The latter constitute the adequate frame for the resolution of several equations used in applications (cf. Section 4.5).

The convolution product is a central theme in distribution theory. It is unavoidable. But the treatment (long or short) differs from a book to another, according to the targets of the authors. Here we tried to make the presentation as accessible as possible without any loss of fundamental results. We put a particular attention in exhibiting the convolution algebras $\mathcal{E}'(\mathbb{R}^n)$ and $\mathcal{D}'_+(\mathbb{R}^n)$. They are sufficient for the treatment of classical partial differential equations. More in [3], [4], [5], [14] and [15].

Chapter 5

Fourier Transformation in $L^1(\mathbb{R}^n)$

We begin by recalling the first properties of Fourier transformation in $L^1(\mathbb{R}^n)$ (algebraic properties and Riemann-Lebesgue theorem). Next we examine the continuity and derivability, giving also the transfer theorem and the preparatory formula of Riesz. Finally we present the inversion theorem and the reciprocity theorem of Fourier-Dirichlet.

5.1 Definitions and first properties.

In the sequel we consider the usual scalar product in $\mathbb{R}^n$, i.e., for $x = (x_1, \ldots, x_n)$ and $y = (y_1, \ldots, y_n)$, we put $xy = x_1 y_1 + \cdots + x_n y_n$. Let $f : \mathbb{R}^n \longrightarrow \mathbb{C}$. If $f \in L^1(\mathbb{R}^n)$ then for every y, the functions $x \longmapsto f(x)e^{-2i\pi xy}$ and $x \longmapsto f(x)e^{2i\pi xy}$ are integrable.

Definition 5.1. For every $f \in L^1(\mathbb{R}^n)$ and every $y \in \mathbb{R}^n$, put

$$\mathcal{F}f(y) = \int_{\mathbb{R}^n} f(x)e^{-2i\pi xy}dx; \ y \in \mathbb{R}^n$$

and

$$\overline{\mathcal{F}}f(y) = \int_{\mathbb{R}^n} f(x)e^{2i\pi xy}dx; \ y \in \mathbb{R}^n.$$

We say that the function $\mathcal{F}f$, denoted also by $\widehat{f}$, is the Fourier transform of f. The function $\overline{\mathcal{F}}f$ is said to be the Fourier cotransform of f.

Remark 5.1.

(i) The maps $\mathcal{F}$ and $\overline{\mathcal{F}}$ are respectively called Fourier transformation and Fourier cotransformation on $\mathbb{R}^n$.

(ii) If $f \in L^1(\mathbb{R}^n)$, then $\mathcal{F}f$ and $\overline{\mathcal{F}}f$ are by definition the images of any representative of f.

Here are some immediate algebraic properties.

Proposition 5.1. *For $f, g \in L^1(\mathbb{R}^n)$ and $\lambda, \mu \in \mathbb{C}$, one has*

(i) $\mathcal{F}(\lambda f + \mu g) = \lambda \mathcal{F}f + \mu \mathcal{F}g$.

(ii) $\overline{\mathcal{F}}(\bar{f}) = \overline{\mathcal{F}f}$.

(iii) $\mathcal{F}\overset{\vee}{f} = \overline{\mathcal{F}}f$.

(iv) $\overline{\mathcal{F}}\overset{\vee}{f} = \mathcal{F}f$.

(v) $\mathcal{F}(\tau_a f) = \chi_{-a}(\mathcal{F}f)$ and $\mathcal{F}(\chi_a f) = \tau_a(\mathcal{F}f)$, $a \in \mathbb{R}$; where $\tau_a f(x) = f(x-a)$ and $\chi_a(x) = e^{2i\pi ax}$.

(vi) $\mathcal{F}(f_1 \otimes \cdots \otimes f_n) = \mathcal{F}(f_1) \otimes \cdots \otimes \mathcal{F}(f_n)$; $f_1, \ldots, f_n \in L^1(\mathbb{R})$.

(vii) $\mathcal{F}(f * g) = \mathcal{F}(f)\, \mathcal{F}(g)$.

The following example is fundamental.

Example 5.1. Let f be the function defined on $\mathbb{R}^n$ by $f(x) = e^{-\pi x^2}$. Then $\mathcal{F}f = \overline{\mathcal{F}}f = f$. Indeed the first equality results from (iv) of the previous proposition, taking into account that $\overset{\vee}{f} = f$. To show that $\mathcal{F}f = f$, this is reduced to $n = 1$, by (vi) of the previous proposition. For every $y \in \mathbb{R}$, one has

$$\mathcal{F}f(y) = \int_{\mathbb{R}} e^{-\pi x^2} e^{-2i\pi xy}\, dx = e^{-\pi y^2} \int_{\mathbb{R}} e^{-\pi(x+iy)^2}\, dx.$$

Consider the complex function $z \longmapsto e^{-\pi z^2}$. It is holomorphic on $\mathbb{C}$. Hence by Cauchy's theorem, its integral along the rectangle with vertices $-a$, a, $a+ib$, and $-a+ib$ is zero. Moreover, its integral on the vertical sides tends to zero when a tends to $+\infty$; for example

$$\left| \int_a^{a+ib} e^{-\pi z^2}\, dz \right| = \left| i \int_0^b e^{-\pi(a^2-y^2)} e^{-2i\pi ay}\, dy \right| \leq |b|\, e^{-\pi(a^2-b^2)}.$$

We then have

$$\int_{\mathbb{R}} e^{-\pi(x+iy)^2}\, dx = \lim_{a \to +\infty} \int_{-a+ib}^{a+ib} e^{-\pi z^2}\, dz = \lim_{a \to +\infty} \int_{-a}^{a} e^{-\pi z^2}\, dz$$

$$= \int_{-\infty}^{+\infty} e^{-\pi t^2}\, dt,$$

and it is known that the last integral equals 1 (Gauss integral).

We now look at the continuity and the derivability of $\mathcal{F}f$.

Theorem 5.1. *Let $f \in L^1(\mathbb{R}^n)$. Then*

(i) $\mathcal{F}f$ is continuous and bounded. Moreover

$$\|\mathcal{F}f\|_\infty \le \|f\|_1.$$

(ii) Riemann-Lebesgue theorem

$$\mathcal{F}f(y) \underset{\|y\|\to+\infty}{\longrightarrow} 0.$$

We say that $\mathcal{F}f$ tends to 0 at infinity.

Proof.

(i) As $\left|f(x)e^{-2i\pi xy}\right| = |f(x)|$ with $f \in L^1(\mathbb{R}^n)$ and $y \longmapsto f(x)e^{-2i\pi xy}$ is continuous *a.e.*, for every x, Lebesgue's theorem shows that $\mathcal{F}f$ is continuous. Moreover

$$|\mathcal{F}f(y)| \le \int_{\mathbb{R}^n} |f(x)|\,dx, \text{ for every } y \in \mathbb{R}^n.$$

(ii) The density of $\mathcal{D}(\mathbb{R}^n)$ in $L^1(\mathbb{R}^n)$ allows to consider $f \in \mathcal{D}(\mathbb{R}^n)$. Indeed for a given $f \in L^1(\mathbb{R}^n)$, there is a sequence $(f_k)_k$ in $\mathcal{D}(\mathbb{R}^n)$ such that $\|f_k - f\|_1 \underset{k}{\longrightarrow} 0$. Then $\|\mathcal{F}f_k - \mathcal{F}f\|_\infty \underset{k}{\longrightarrow} 0$ by the inequality of (i). Let now $f \in \mathcal{D}(\mathbb{R}^n)$. Using Fubini's theorem and an integration by parts, one has

$$\mathcal{F}f(y) = \frac{1}{2\pi i y_j}\mathcal{F}\left(\frac{\partial f}{\partial x_j}\right)(y); \ j = 1, 2, ..., n \text{ and } y_j \ne 0.$$

Then

$$|\mathcal{F}f(y)| \le \frac{1}{2\pi |y_j|}\left\|\frac{\partial f}{\partial x_j}\right\|_1; \ j = 1, 2, ..., n \text{ and } y_j \ne 0.$$

Whence the result. $\qquad\square$

Concerning results on the derivation, we recall first a useful function. Designate by M^α, $\alpha = (\alpha_1, ..., \alpha_n) \in \mathbb{N}^n$, the function from $\mathbb{R}^n$ into $\mathbb{R}$ defined by

$$M^\alpha : x \longmapsto x^\alpha = x_1^{\alpha_1} \ldots x_n^{\alpha_n}.$$

It is said to be the monomial multiplication.

Theorem 5.2. *Let $f \in L^1(\mathbb{R}^n)$ and $k \in \mathbb{N}$.*

(i) *If $M^\alpha f \in L^1(\mathbb{R}^n)$ for every $\alpha \in \mathbb{N}^n$ with $|\alpha| \le k$, then $\hat{f} \in C^k(\mathbb{R}^n)$ and we have*

$$D^\alpha \hat{f} = \widehat{(-2i\pi M)^\alpha f}.$$

(ii) *If $f \in C^k(\mathbb{R}^n)$ is such that $D^\beta f \in L^1(\mathbb{R}^n)$ for every $\beta \in \mathbb{N}^n$ with $|\beta| \le k$, then*

$$M^\beta \hat{f} \in L^\infty(\mathbb{R}^n) \text{ and } (2\pi i M)^\beta \hat{f} = \widehat{D^\beta f}.$$

Proof.

(i) It is sufficient to show the result for $|\alpha| = 1$. Take for example $\alpha = (1, 0, ..., 0)$. By hypothesis, the function $x \longmapsto x_1 f(x)$ is integrable on $\mathbb{R}$. We will show that $\hat{f}$ is differentiable with respect to y_1 and that

$$\frac{\partial \hat{f}}{\partial y_1}(y) = \int_{\mathbb{R}^n} -2\pi i x_1 f(x)\, e^{-2\pi i x y} dx.$$

Since the function $y \longmapsto -2i\pi x_1 e^{-2i\pi x y} f(x)$ is continuous on $\mathbb{R}^n$ for almost every $x \in \mathbb{R}^n$ and

$$\left| -2i\pi x_1 f(x) e^{-2i\pi x y} \right| = 2\pi |x_1 f(x)|,$$

one can differentiate under the integral sign.

(ii) It is sufficient to show that if $\frac{\partial f}{\partial x_1} \in C(\mathbb{R}^n) \cap L^1(\mathbb{R}^n)$, then

$$2i\pi y_1 \hat{f}(y) = \int_{\mathbb{R}^n} \frac{\partial f}{\partial x_1}(x) e^{-2i\pi x y} dx,$$

i.e.,

$$\int_{\mathbb{R}^n} \left(\frac{\partial f}{\partial x_1}(x) e^{-2i\pi x y} - 2i\pi y_1 f(x) e^{-2i\pi x y} \right) dx = 0,$$

or yet

$$\int_{\mathbb{R}^n} \frac{\partial}{\partial x_1} \left[f(x) e^{-2i\pi x y} \right] dx = 0.$$

This is equivalent, by Fubini's theorem, to

$$\int_{\mathbb{R}^n} \left[f(x) e^{-2i\pi x y} \right]_{x_1 = -\infty}^{x_1 = +\infty} dx_2 \cdots dx_n = 0.$$

But

$$\left[f(x) e^{-2i\pi x y} \right]_{x_1 = -\infty}^{x_1 = +\infty} = 0 \text{ for } \lim_{|x| \to +\infty} f(x) = 0,$$

knowing that f and $\frac{\partial f}{\partial x_1}$ are in $C(\mathbb{R}^n) \cap L^1(\mathbb{R}^n)$. $\qquad\square$

The following results (the transfer theorem and the preparatory formula of Riesz) will be useful in the sequel.

Theorem 5.3 (Transfer Theorem). *Let $f,\, g \in L^1(\mathbb{R}^n)$. Then*

$$\int_{\mathbb{R}^n} f(x)\widehat{g}(x)dx = \int_{\mathbb{R}^n} \widehat{f}(y)g(y)dy.$$

Proof. First the functions $f\widehat{g}$ and $\widehat{f}g$ are in $L^1(\mathbb{R}^n)$ for $|f\widehat{g}| \le \|g\|_1 |f|$ and $\left|\widehat{f}g\right| \le \|f\|_1 |g|$. Now using repeatedly Fubini's theorem, one has

$$\int_{\mathbb{R}^n} f(x)\widehat{g}(x)dx = \int_{\mathbb{R}^n} f(x) \left(\int_{\mathbb{R}^n} g(y)e^{-2i\pi xy}dy \right) dx$$

$$= \int_{\mathbb{R}^n} \int_{\mathbb{R}^n} f(x)g(y)e^{-2i\pi xy}dydx$$

$$= \int_{\mathbb{R}^n} g(y) \left(\int_{\mathbb{R}^n} f(x)e^{-2i\pi xy}dx \right) dy$$

$$= \int_{\mathbb{R}^n} g(y)\hat{f}(y)dy.$$

$\square$

Using the first delay formula (cf. (v) of Proposition 5.1), the transfer theorem, applied to the functions $\tau_{-a}f$, $a \in R^n$ and g, one has

Corollary 5.1 (Preparatory Formula of Riesz). *Let $f,\, g \in L^1(\mathbb{R}^n)$ and $a \in \mathbb{R}^n$. Then*

$$\int_{\mathbb{R}^n} f(a+x)\hat{g}(x)dx = \int_{\mathbb{R}^n} \hat{f}(y)g(y)e^{2i\pi ay}dy.$$

5.2 Inversion Theorem

If $f \in L^1(\mathbb{R})$, let $(c_n)_n$ be the sequence of its Fourier coefficients given by

$$c_n = \int_{\mathbb{R}} f(t)e^{-2i\pi nt}dt.$$

One of the questions concerns the reconstruction of the function f from its Fourier series, i.e., to have $f(x) = \sum_{n=-\infty}^{+\infty} c_n e^{2i\pi nx}$. This equality, when it takes place, is called the inversion formula. If instead of the sequence

$(c_n)_n$, one considers the Fourier transformation $\hat{f}$ of f, the analogue of the inversion formula is

$$f(x) = \int_{\mathbb{R}^n} \hat{f}(y) e^{2i\pi xy} dy.$$

The right-hand side has a meaning, in particular whenever $\hat{f} \in L^1(\mathbb{R})$. But the latter condition is not always fulfilled for if we take for example

$$f(x) = \begin{cases} 1 \text{ if } |x| \leq 1 \\ 0 \text{ if } |x| > 1, \end{cases}$$

then

$$\hat{f}(y) = \frac{\sin 2\pi y}{\pi y}.$$

The right-hand side of the last inversion formula has the same value for any function equal to f almost everywhere. We must thus expect an equality only almost everywhere.

Proposition 5.2 (Inversion Theorem). *If $f \in L^1(\mathbb{R}^n)$ and $\widehat{f} \in L^1(\mathbb{R}^n)$, then*

$$f(x) = \int_{\mathbb{R}^n} \hat{f}(y) e^{2i\pi xy} dy, \text{ for almost every } x \in \mathbb{R}^n.$$

Proof. One has

$$\int_{\mathbb{R}^n} \widehat{f}(y) e^{-2i\pi xy} dy = \int_{\mathbb{R}^n} \left(\int_{\mathbb{R}^n} f(t) e^{-2i\pi ty} dt \right) e^{2i\pi xy} dy.$$

This reminds us of Fubini's theorem. But the function

$$(y,t) \longmapsto f(t) e^{2i\pi y(x-t)}$$

is never integrable on $\mathbb{R} \times \mathbb{R}$ for a non zero f, due to $\left| f(t) e^{2i\pi y(x-t)} \right| = |f(t)|$. However it is so on $K \times \mathbb{R}$ for every compact subset K in $\mathbb{R}$. Taking $K = [-p,p]^n$, $p \in \mathbb{N}$, one remarks that

$$\int_{\mathbb{R}^n} \widehat{f}(y) e^{2i\pi xy} dy = \lim_{p \to +\infty} \int_{[-p,p]^n} \left(\int_{\mathbb{R}^n} f(t) e^{-2i\pi ty} dt \right) e^{2i\pi xy} dy$$

since $\widehat{f} \in L^1(\mathbb{R}^n)$ by hypothesis, and

$$\int_{[-p,p]^n} \left(\int_{\mathbb{R}^n} f(t) e^{-2i\pi ty} dt \right) e^{2i\pi xy} dy = \int_{\mathbb{R}^n} \int_{[-p,p]^n} f(t) e^{2i\pi y(x-t)} dt dy$$

$$= \int_{\mathbb{R}^n} \int_{[-p,p]^n} f(x-u) e^{-2i\pi yu} du dy$$

$$= \int_{\mathbb{R}^n} \left(\int_{[-p,p]^n} e^{2i\pi yu} dy \right) f(x-u) du$$

$$= \int_{\mathbb{R}^n} f_p(u) g(u) du,$$

where

$$f_p(u) = \int_{[-p,p]^n} e^{2i\pi yu} dy \text{ and } g(u) = f(x-u).$$

According 3) of Examples 2.3, Section 2.8 of Chapter 2, one has for every $f \in \mathcal{D}(\mathbb{R}^n)$

$$\lim_{p \to +\infty} \int_{\mathbb{R}^n} f_p(u)g(u)du = g(0) = f(x).$$

Thus the equality takes place in $\mathcal{D}(\mathbb{R}^n)$. Hence it is valid for the regularizations of f, i.e.,

$$f * \theta_j(x) = \int_{\mathbb{R}^n} \widehat{f * \theta_j}(y)e^{2i\pi xy}dy; \qquad x \in \mathbb{R}^n$$
$$= \int_{\mathbb{R}^n} \hat{f}(y)\overset{\wedge}{\theta_j}(y)e^{2i\pi xy}dy.$$

But for $y \in \mathbb{R}^n$

$$\left| \hat{f}(y)\overset{\wedge}{\theta_j}(y)e^{2i\pi xy} \right| \le \left| \hat{f}(y) \right| \text{ and } \lim_{j \to +\infty} \hat{f}(y)\overset{\wedge}{\theta_j}(y)e^{2i\pi xy} = \hat{f}(y)e^{2i\pi xy},$$

since

$$\left| \overset{\wedge}{\theta_j}(y) - 1 \right| \le \sup_{\|x\| \le \varepsilon_j} \left| e^{-2i\pi xy} - 1 \right| \text{ with } \varepsilon_j \to 0.$$

By the Lebesgue theorem, one has

$$f * \theta_j(x) \xrightarrow[j \to +\infty]{} \int_{\mathbb{R}^n} \hat{f}(y)e^{2i\pi xy}dy.$$

Finally, since $(f * \theta_j)_j$ converges to f in $L^1(\mathbb{R}^n)$, there is a subsequence of $(f * \theta_j)_j$ which converges to f almost everywhere, so that we have

$$f(x) = \int_{\mathbb{R}^n} \hat{f}(y)e^{2i\pi xy}dy, \text{ almost everywhere.}$$

$\square$

An immediate consequence is the following.

Corollary 5.2 (Uniqueness theorem). *Let* $f \in L^1(\mathbb{R}^n)$ *such that* $\hat{f} = 0$ *on* $\mathbb{R}^n$. *Then* $f = 0$ *almost everywhere.*

If $f \in L^1(\mathbb{R}^n)$ and $\hat{f} \in L^1(\mathbb{R}^n)$, the inversion formula shows that f is continuous almost everywhere. So the pointwise reconstruction of f is possible only for functions which are almost everywhere continuous. Functions of bounded variation (cf. [11]) are almost everywhere continuous. In this case, for $n = 1$ and $f \in L^1(\mathbb{R})$ one obtains the analog of Dirichlet's inversion formula. If $f \in L^1(\mathbb{R})$ and $p \in \mathbb{N}$, then one has by Fubini's theorem

$$\int_{-p}^{p} \hat{f}(y) e^{2i\pi xy} dy = \frac{1}{\pi} \int_{-\infty}^{+\infty} f(x - u) \frac{\sin 2\pi pu}{u} du.$$

This leads, as in the classical case of series, to show the following technical results.

Proposition 5.3. *Let f be a function with bounded variation on any interval $[a, b]$ in $\mathbb{R}$. Then*

(i) Riemann-Lebesgue lemma:

$$\lim_{\lambda \to +\infty} \int_{a}^{b} f(t) \sin \lambda t \, dt = 0.$$

(ii) Dirichlet's lemma: For every $a > 0$, one has

$$\lim_{\lambda \to +\infty} \int_{0}^{a} f(t) \frac{\sin \lambda t}{t} dt = \frac{\pi}{2} f(0^+).$$

Proof.

(i) We may consider only an increasing function, for any function with bounded variation is the difference of two such functions. Then by the second mean value theorem, there is $c \in [a, b]$ such that

$$\int_{a}^{b} f(t) \sin \lambda t \, dt = f(a) \int_{a}^{c} \sin \lambda t \, dt + f(b) \int_{c}^{b} \sin \lambda t \, dt,$$

whence

$$\left| \int_{a}^{b} f(t) \sin \lambda t \, dt \right| \le \frac{2}{\lambda} (|f(a)| + |f(b)|).$$

(ii) We may suppose f decreasing. Then we have

$$\int_{0}^{a} f(t) \frac{\sin \lambda t}{\lambda} dt - f(0^+) \int_{0}^{+\infty} \frac{\sin s}{s} ds$$

$$= \int_{0}^{a} [f(t) - f(0^+)] \frac{\sin \lambda t}{t} dt + f(0^+) \left[\int_{0}^{a} \frac{\sin \lambda t}{\lambda} dt - \int_{0}^{+\infty} \frac{\sin s}{s} ds \right].$$

The second term of the right-hand side tends to zero by the substitution $u = \lambda t$. Now let $\varepsilon > 0$. There is η, $0 < \eta < a$, such that

$$|f(t) - f(0^+)| \le \varepsilon \text{ whenever } 0 \le t \le \eta.$$

Yet, we have

$$\int_0^a [f(t) - f(0^+)] \frac{\sin \lambda t}{t} dt = \int_0^\eta g_\lambda(t) dt + \int_\eta^a g_\lambda(t) dt,$$

where

$$g_\lambda(t) = [f(t) - f(0^+)] \frac{\sin \lambda t}{t}.$$

The function

$$t \longmapsto \frac{f(t) - f(0^+)}{t}$$

is of bounded variation on $[\eta, a]$ as a difference of the two decreasing functions

$$t \longmapsto \frac{f(t) - f(0^+)}{t} \text{ and } t \longmapsto \frac{f(0^+) - f(a)}{t}.$$

By (i), one has

$$\lim_{\lambda \to +\infty} \int_\eta^a g_\lambda(t) \, dt = 0.$$

By the second mean formula, there is $c \in [0, \eta]$ such that

$$\int_o^\eta g_\lambda(t) \, dt$$

$$= [f(\eta) - f(0^+)] \int_c^\eta \frac{\sin \lambda t}{t} \, dt + [f(0) - f(0^+)] \int_0^c \frac{\sin \lambda t}{t} \, dt,$$

and we have

$$\left| \int_0^\eta g_\lambda(t) \, dt \right| \le \varepsilon \left[\left| \int_c^\eta \frac{\sin \lambda t}{t} \, dt \right| + \left| \int_0^c \frac{\sin \lambda t}{t} \, dt \right| \right] \le 2\varepsilon k,$$

where

$$k = \sup \left\{ \left| \int_u^{u'} \frac{\sin t}{t} dt \right| ; \ 0 \le u, \ u \le u' \right\}$$

is finite due to the convergence of $\int_0^{+\infty} \frac{\sin t}{t} \, dt$ which is equal to $\frac{\pi}{2}$.

$\square$

Now, here is the analogue of the Dirichlet inversion formula:

Proposition 5.4 (Fourier-Dirichlet reciprocity theorem). *If $f \in L^1(\mathbb{R})$ is of locally bounded variation (i.e., with bounded variation on every compact interval of $\mathbb{R}$), then for every $x \in \mathbb{R}$,*

$$\lim_{p \to +\infty} \int_{-p}^{p} \hat{f}(y) e^{2i\pi xy} dy = \frac{1}{2} \left(f(x^+) + f(x^-) \right); \ p \in \mathbb{N}.$$

Proof. For any $p \in \mathbb{N}^*$, one has

$$\int_{-p}^{p} \hat{f}(y) e^{2i\pi xy} dy = \frac{1}{\pi} \int_{-\infty}^{+\infty} f(x-u) \frac{\sin 2\pi pu}{u} du.$$

By the Dirichlet formula ((ii), Proposition 5.3), one gets for every $a > 0$

$$\lim_{p \to +\infty} \int_{0}^{a} f(x-u) \frac{\sin 2\pi pu}{u} \, du = \frac{\pi}{2} f(x^-)$$

and

$$\lim_{p \to +\infty} \int_{-a}^{0} f(x-u) \frac{\sin 2\pi pu}{u} du = \frac{\pi}{2} f(x^+).$$

On the other hand

$$\lim_{p \to +\infty} \int_{|u| \geq a} f(x-u) \frac{\sin 2\pi pu}{u} du = 0,$$

since

$$\int_{|u| \geq a} \left| \frac{f(x-u)}{u} \right| du \leq \frac{1}{a} \|f\|_1 \, .$$

We then obtain the result. $\qquad \square$

Remark 5.2. We have used

$$\int_{0}^{+\infty} \frac{\sin s}{s} ds = \frac{\pi}{2}.$$

Actually, we can argue without knowing this value a priori and even find it again. Indeed, the previous calculations give

$$\lim_{p \to +\infty} \int_{-p}^{p} \hat{f}(y) e^{2i\pi xy} dy = \left(\frac{2}{\pi} \int_{0}^{+\infty} \frac{\sin s}{s} ds \right) \frac{f(x^+) + f(x^-)}{2}.$$

Taking a non zero continuous function satisfying the hypothesis of Proposition 5.2 e.g., $x \longmapsto e^{-\pi x^2}$, one obtains for every x

$$f(x) = \left(\frac{2}{\pi} \int_0^{+\infty} \frac{\sin s}{s} ds \right) f(x).$$

It is well known that in the convolution algebra $L^1(\mathbb{R}^n)$ the Fourier transformation $\mathcal{F}$ plays a very important role. The same is true in the theory of distributions. One can say that the space $L^1(\mathbb{R}^n)$ is the test space for the Fourier transformation. This chapter is of a general interest as it was the case with the first one. It concerns functions without any link with distributions; so it can be useful for people working in other branches of mathematics, as well.

As a pedagogical choice, this chapter is devoted to fundamental properties of $\mathcal{F}$ in $L^1(\mathbb{R}^n)$ to avoid confusion with other notions, fixing those properties for later extensions.

We follow here in the presentation, more or less, [15]. However there is a difference in conducting the proofs. Concerning the important inversion theorem for example, we prefer to follow [3] where the result is established in the test space $\mathcal{D}(\mathbb{R}^n)$. We complete the proof, using the regularization method; this procedure seems to be more natural. In [6] and [15], the proof relies, from the beginning, on the preparatory formula of Riesz. We note here that in the other references the results are considered in more general setting. More one can find in [3], [4], [5], [9], [12] and [15].

Chapter 6

Tempered Distributions

In this chapter, we first present another space which plays an essential role in the theory of distributions. This is the space $\mathcal{S}(\mathbb{R}^n)$ of functions of class $\mathcal{C}^\infty$ which are rapidly decreasing as well as their derivatives of any order. This space contains $\mathcal{D}(\mathbb{R}^n)$ and is contained in every $L^p(\mathbb{R}^n)$, $p \geq 1$. Several classical operations behave well in this space (derivation, ordinary product, convolution product, multiplication by polynomials,...). The space $\mathcal{S}(\mathbb{R}^n)$ is not stable for the multiplication by functions of class $\mathcal{C}^\infty$, which leads to the notion of a multiplication operator. Fourier transformation is a topological isomorphism from $\mathcal{S}(\mathbb{R}^n)$ onto $\mathcal{S}(\mathbb{R}^n)$. We also deal with general properties of tempered distributions which are the elements of the topological dual $\mathcal{S}'(\mathbb{R}^n)$ of $\mathcal{S}(\mathbb{R}^n)$. The convolution product of a tempered distribution with a function in $\mathcal{S}(\mathbb{R}^n)$ is not necessarily a regular distribution given by a function in $\mathcal{S}(\mathbb{R}^n)$, which leads to the notion of a convolution operator. Fourier transformation in $\mathcal{S}'(\mathbb{R}^n)$ extends that defined in $L^1(\mathbb{R}^n)$ and has analogous properties. In order to extend Fourier transformation to the space $\mathcal{D}'(\mathbb{R}^n)$, of all distributions, we are led to consider the notion of an ultradistribution.

6.1 The space $\mathcal{S}(\mathbb{R}^n)$

If $f \in L^1(\mathbb{R})$, one has

$$\hat{f}(y) = \int_{\mathbb{R}} f(x)\, e^{-2i\pi xy} dx.$$

If f is continuous with compact support, one can write

$$\hat{f}(y) = \left\langle T_f,\ e^{-2i\pi xy} \right\rangle.$$

In fact $\langle T, \, e^{-2i\pi xy} \rangle$ has a meaning whenever T has compact support and it is a function in y. This allows to define the Fourier transform of any distribution with compact support as a function, by

$$\hat{T}(y) = \langle T, \, e^{-2i\pi xy} \rangle.$$

Evidently, this is not possible for an arbitrary distribution. As usual, one first examines the case of regular distributions. For $f \in L^1(\mathbb{R})$ it is natural to define $\hat{T}_f$ as $T_{\hat{f}}$. Hence one must have

$$\left\langle \hat{T}_f, \varphi \right\rangle = \left\langle T_{\hat{f}}, \varphi \right\rangle = \int_{\mathbb{R}} \hat{f}(x)\varphi(x)dx = \int_{\mathbb{R}} f(x)\overset{\wedge}{\varphi}(x)dx,$$

where the last equality results from the transfer formula. So $\left\langle \hat{T}_f, \varphi \right\rangle = \left\langle T_f, \overset{\wedge}{\varphi} \right\rangle$ provided that $\left\langle T_f, \overset{\wedge}{\varphi} \right\rangle$ is defined. But $\overset{\wedge}{\varphi}$ has never compact support. Indeed, one has for every $y \in \mathbb{R}$

$$\overset{\wedge}{\varphi}(y) = \int_{\mathbb{R}} \varphi(x)e^{-2i\pi xy}dx.$$

It admits a series expansion in the whole of $\mathbb{R}$. Hence it is of a compact support only if it is identically zero. This leads us to look for a subspace of $\mathcal{E}(\Omega)$ containing $\mathcal{D}(\Omega)$ and stable by the Fourier transformation. Theorem 5.2 of Chapter 5 suggests a space to consider. Put

$$S\left(\mathbb{R}^n\right) = \left\{ f \in \mathcal{E}\left(\mathbb{R}^n\right) : M^\alpha D^\beta f \in \mathcal{L}^1\left(\mathbb{R}^n\right), \, \forall \alpha, \beta \in \mathbb{N}^n \right\}.$$

It is a vector space such that $\mathcal{D}\left(\mathbb{R}^n\right) \subset S\left(\mathbb{R}^n\right) \subset \mathcal{E}\left(\mathbb{R}^n\right)$.

Remark 6.1. A function $f \in \mathcal{E}\left(\mathbb{R}^n\right)$ is in $S\left(\mathbb{R}^n\right)$ if and only if

$$\left(M^\alpha D^\beta f\right)(x) \underset{|x| \to +\infty}{\longrightarrow} 0, \text{ for every } \alpha, \beta \in \mathbb{N}^n.$$

Necessity is due to the fact that any function of class $\mathcal{C}^1$ which is also integrable tends to zero at infinity. For the converse, take the multi-index $(\alpha_1 + 2, ..., \alpha_n + 2)$. A function $f \in \mathcal{E}\left(\mathbb{R}^n\right)$ such that $\left(M^\alpha D^\beta f\right)(x) \underset{|x| \to +\infty}{\longrightarrow} 0$ for every $\alpha, \beta \in \mathbb{N}^n$, is said to be rapidly decreasing (or of rapid decrease) together with its derivatives for any order.

In the study of $S\left(\mathbb{R}^n\right)$, the polynomials $x \longmapsto M^\alpha(x) = x^\alpha$ and $x \longmapsto \left(1 + \|x\|^2\right)^k$, $k \in \mathbb{N}$, play an important role. They are linked by the following lemma.

Lemma 6.1. *For every $k \in \mathbb{N}$, one has*

(i) $\left|x^{2\alpha}\right| \leq \left(1 + \|x\|^2\right)^k$; $\forall x \in \mathbb{R}$, $\forall \alpha \in \mathbb{N}^n$, $|\alpha| \leq k$.

(ii) There is $c > 0$ such that $\left(1 + \|x\|^2\right)^k \leq c \sup_{|\alpha| \leq k} \left|x^{2\alpha}\right|$.

Proof.

(i) One has
$$\left|x^{2\alpha}\right| = \left(|x_1|^2\right)^{\alpha_1} \cdots \left(|x_n|^2\right)^{\alpha_n} \leq \left(1 + \|x\|^2\right)^k.$$

(ii) Results from the following Newton multinomial formula
$$\left(1 + x_1^2 + \cdots + x_n^2\right)^k = \sum_{|\alpha| \leq k} \frac{k!}{\alpha!\,(k - |\alpha|)!} x_1^{2\alpha_1} \ldots x_n^{2\alpha_n}.$$

□

Here are characterizations of the space $\mathcal{S}\left(\mathbb{R}^n\right)$.

Proposition 6.1. *For $f \in \mathcal{E}\left(\mathbb{R}^n\right)$, the following assertions are equivalent.*

(i) $f \in \mathcal{S}\left(\mathbb{R}^n\right)$.

(ii) $\left\|M^\alpha D^\beta f\right\|_\infty < +\infty$; $\forall \alpha, \beta \in \mathbb{N}^n$.

(iii) $\sup_{x \in \mathbb{R}^n} \left|\left(1 + \|x\|^2\right)^k D^\beta f(x)\right| < +\infty$, $\forall\, k \in \mathbb{N}$, $\forall \beta \in \mathbb{N}^n$.

Proof. (i)$\Longrightarrow$(ii) By Remark 6.1.

(ii)$\Longrightarrow$(iii) By (ii) of Lemma 6.1.

(iii)$\Longrightarrow$(i) For every $k \in \mathbb{N}$, one has
$$x^\alpha D^\beta f(x) = \frac{x^\alpha}{\left(1 + \|x\|^2\right)^k} \left(1 + \|x\|^2\right)^k D^\beta f(x).$$

By (iii), the function $x \longmapsto \left(1 + \|x\|^2\right)^k D^\beta f(x)$ is bounded for every k. It is sufficient to take k such that
$$x \longmapsto \frac{x^\alpha}{\left(1 + \|x\|^2\right)^k}$$

be integrable.

□

For every $f \in \mathcal{S}(R^n)$ and $\alpha, \beta \in \mathbb{N}^n$, put
$$p_{\alpha,\beta}(f) = \sup_{x \in \mathbb{R}^n} \left|x^\alpha D^\beta f(x)\right|.$$

The $p_{\alpha,\beta}$'s are seminorms on $\mathcal{S}(\mathbb{R}^n)$. They determine a topology of a separated locally convex space on $\mathcal{S}(\mathbb{R}^n)$. It is said to be the natural topology of $\mathcal{S}(\mathbb{R}^n)$. For every $f \in \mathcal{S}(\mathbb{R}^n)$, $k \in \mathbb{N}$ and $\alpha, \beta \in \mathbb{N}^n$, put also

$$q_{k,\beta}(f) = \sup_{x \in \mathbb{R}^n} \left| \left(1 + \|x\|^2\right)^k D^\beta f(x) \right|$$

and

$$r_{\alpha,\beta}(f) = \left\| M^\alpha D^\beta f \right\|_1.$$

One still obtains seminorms on $\mathcal{S}(\mathbb{R}^n)$. That is, we have the following result.

Proposition 6.2. *Each of the families* $(q_{k,\beta})_{k,\beta}$ *and* $(r_{\alpha,\beta})_{\alpha,\beta}$ *of seminorms defines the natural topology of* $\mathcal{S}(\mathbb{R}^n)$.

Proof. Using Lemma 6.1 one shows that $q_{k,\beta} \le c p_{\alpha',\beta}$, where $\alpha' = (2k, ..., 2k)$, and that $p_{\alpha,\beta} \le q_{n|\alpha|,\beta}$. Hence $(q_{k,\beta})_{k,\beta}$ defines the natural topology of $\mathcal{S}(\mathbb{R}^n)$. On the other hand for any $\alpha, \beta \in \mathbb{N}^n$ and a suitable k, one has

$$r_{\alpha,\beta} \le c_\alpha q_{k,\beta}, \text{ where } c_\alpha = \int_{\mathbb{R}^n} \frac{x^\alpha}{\left(1 + \|x\|^2\right)^k} dx.$$

Now since $q_{k,\beta} \le c p_{\alpha,\beta}$ with $\alpha = (2k, ..., 2k)$, it remains to show that any neighborhood of zero for the topology given by the $p_{\alpha,\beta}$'s is a neighborhood of zero for that one defined by the $r_{\alpha,\beta}$'s. One has

$$p_{\alpha,\beta}(f) = \sup_{x \in \mathbb{R}^n} \left| x^\alpha D^\beta f(x) \right| = p_{0,0} \left(M^\alpha D^\beta f \right).$$

Put $g_{\alpha,\beta} = M^\alpha D^\beta f$. Since $f \in \mathcal{S}(\mathbb{R}^n)$, the function $g_{\alpha,\beta}$ tends to zero at infinity. Hence

$$g_{\alpha,\beta}(x) = \int_{-\infty}^{x_1} \cdots \int_{-\infty}^{x_n} \frac{\partial^n}{\partial x_1 \cdots \partial x_n} g_{\alpha,\beta}(x) dx.$$

That gives

$$p_{0,0}(g_{\alpha,\beta}) \le r_{0,\gamma}(g_{\alpha,\beta}) = \left\| D^\gamma (M^\alpha D^\beta f) \right\|_1; \text{ where } \gamma = (1, \ldots, 1).$$

One then easily checks that there is $c_\alpha > 0$ such that

$$\left\| D^\gamma (M^\alpha D^\beta f) \right\|_1 \le c_\alpha \max_{|\beta'| \le |\beta|+n} \left\| M^{\alpha'} D^{\beta'} f \right\|_1; \text{ where } |\alpha'| = |\alpha| - 1.$$

Finally

$$p_{\alpha,\beta}(f) \le c_\alpha \max_{|\beta'| \le |\beta|+n} r_{\alpha',\beta'}(f).$$

$\square$

The natural topology of $\mathcal{S}(\mathbb{R}^n)$ is also given by the family $(p_{l,m})_{l,m\in\mathbb{N}}$ of seminorms defined by

$$p_{l,m} = \max\{p_{\alpha,\beta} : |\alpha| \le l, |\beta| \le m\}.$$

Hence it is metrizable. In effect, we have something more. Indeed, a similar proof to that one of Proposition 1.2 of Chapter 1 supplies the following result.

Proposition 6.3. $(\mathcal{S}(\mathbb{R}^n), (p_{\alpha,\beta})_{\alpha,\beta})$ *is a Fréchet space.*

We now examine the stability of the space $\mathcal{S}(\mathbb{R}^n)$ for the usual and the convolution products. Using the Leibniz formula, one sees that it is stable for the usual product. Moreover the bilinear map $(f,g) \longmapsto fg$, from $\mathcal{S}(\mathbb{R}^n) \times \mathcal{S}(\mathbb{R}^n)$ into $\mathcal{S}(\mathbb{R}^n)$, is continuous since there is $c_m > 0$ such that for any f,g in $\mathcal{S}(\mathbb{R}^n)$

$$p_{\alpha,m}(fg) \le c_m p_{0,m}(f)p_{\alpha,m}(g); \ \alpha \in \mathbb{N}^n, \ m \in \mathbb{N}.$$

Concerning the convolution product one has the following result.

Proposition 6.4. *The space* $\mathcal{S}(\mathbb{R}^n)$ *is stable for the convolution product and the bilinear mapping* $(f,g) \longmapsto f * g$ *is continuous from* $\mathcal{S}(\mathbb{R}^n) \times \mathcal{S}(\mathbb{R}^n)$ *into* $\mathcal{S}(\mathbb{R}^n)$.

Proof. Let us first show the stability for the convolution product. Let $f,g \in \mathcal{S}(\mathbb{R}^n)$. Since $f,g \in \mathcal{C}^\infty(\mathbb{R}^n) \cap L^1(\mathbb{R}^n)$, one has

$$D^\beta(f * g) = (D^\beta f) * g = f * D^\beta g; \ \forall \beta \in \mathbb{N}^n.$$

And as $D^\beta f \in \mathcal{S}(\mathbb{R}^n)$, it is sufficient to show that $M^\alpha(f * g)$ is in $L^1(\mathbb{R}^n)$. But this ensues from the following formula

$$M^\alpha(f * g) = \sum_{\delta \le \alpha} \frac{\alpha!}{(\alpha - \delta)!\delta!} M^{\alpha-\delta} f * M^\delta g;$$

an indication on calculations is

$$
\begin{aligned}
M^\alpha(f * g) &= \int_{\mathbb{R}^n} x^\alpha f(x-y)g(y)dy \\
&= \int_{\mathbb{R}^n} \sum_{\delta \le \alpha} \frac{\alpha!}{(\alpha-\delta)!\delta!}(x-y)^{\alpha-\delta}y^\delta f(x-y)g(y)dy \\
&= \sum_{\delta \le \alpha} \frac{\alpha!}{(\alpha-\delta)!\delta!}(M^{\alpha-\delta}f * M^\delta g)(x).
\end{aligned}
$$

Using the previous formula, applied to $D^\beta f$ and g, one obtains

$$r_{l,\beta}(f * g) \le c_l \, r_{l,\beta}(f) \, r_{l,0}(g)$$

where c_l is a positive constant. Whence the continuity of the convolution product.

$\square$

Proposition 6.5. *For any $\gamma \in \mathbb{N}^n$, the linear mappings D^γ and M^γ from $\mathcal{S}(\mathbb{R}^n)$ into $\mathcal{S}(\mathbb{R}^n)$ are continuous.*

Proof. Continuity ensues from the following relations

$$p_{\alpha,\beta}(D^\gamma f) = p_{\alpha,\beta+\gamma}(f) \quad \text{and} \quad p_{l,m}(M^\gamma f) \le c_m p_{l+|\gamma|,\,m}(f)$$

where c_m is a positive constant.

$\square$

Now we will compare $\mathcal{S}(\mathbb{R}^n)$ with respect to the other spaces considered so far.

Proposition 6.6. *The space $\mathcal{S}(\mathbb{R}^n)$ is contained in every $L^p(\mathbb{R}^n)$, $1 \le p \le +\infty$. Moreover the embedding $\mathcal{S}(\mathbb{R}^n) \longrightarrow L^p(\mathbb{R}^n)$ is continuous.*

Proof. We know that $\mathcal{S}(\mathbb{R}^n) \subset L^\infty(\mathbb{R}^n) \cap L^1(\mathbb{R}^n)$. Continuity of the injection is due to $\|f\|_\infty = p_{0,0}(f)$ and $\|f\|_1 = r_{0,0}(f)$. For $1 < p < +\infty$, one has

$$|f(x)|^p \le \frac{1}{\left(1 + \|x\|^2\right)^{np}} \, q_{n,0}(f)^p.$$

The function

$$x \longmapsto \frac{1}{\left(1 + \|x\|^2\right)^{np}}$$

is integrable for

$$\left(1 + \|x\|^2\right)^{np} \ge (1 + x_1^2) \cdots (1 + x_n^2).$$

So $f \in L^1(\mathbb{R}^n)$ and $\|f\|_p \le \pi^{\frac{n}{p}} q_{n,0}(f)$.

$\square$

Recall that $\mathcal{C}_0(\mathbb{R}^n)$ is the space of continuous functions on $\mathbb{R}^n$ which vanish at infinity.

Proposition 6.7.

(i) The injection $\mathcal{S}(\mathbb{R}^n) \longrightarrow \mathcal{C}_0(\mathbb{R}^n)$ is continuous.

(ii) The injection $\mathcal{S}(\mathbb{R}^n) \longrightarrow \mathcal{E}(\mathbb{R}^n)$ is continuous.

Proof.

(i) We have already seen that $\mathcal{S}(\mathbb{R}^n) \subset \mathcal{C}_0(\mathbb{R}^n)$. The continuity of the injection is due to $\|f\|_\infty = p_{0,0}(f)$.

(ii) One has by definition $\mathcal{S}(\mathbb{R}^n) \subset \mathcal{E}(\mathbb{R}^n)$. The continuity of the injection results from the fact that, for every compact K in $\mathbb{R}^n$ and every $m \in \mathbb{N}$,

$$p_{K,m}(f) = \sup\left\{\left|D^\beta f(x)\right| : |\beta| \le m,\ x \in K\right\} \le p_{0,m}(f). \qquad \square$$

Proposition 6.8.

(i) The space $\mathcal{D}(\mathbb{R}^n)$ is contained in $\mathcal{S}(\mathbb{R}^n)$ and the injection $\mathcal{D}(\mathbb{R}^n) \longrightarrow \mathcal{S}(\mathbb{R}^n)$ is continuous.

(ii) $\mathcal{D}(\mathbb{R}^n)$ is dense in $\mathcal{S}(\mathbb{R}^n)$.

Proof.

(i) It is immediate that $\mathcal{D}(\mathbb{R}^n) \subset \mathcal{S}(\mathbb{R}^n)$. The injection is continuous if and only if for every compact K in $\mathbb{R}$, the injection $\mathcal{D}_K(\mathbb{R}^n) \longrightarrow \mathcal{S}(\mathbb{R}^n)$ is continuous. But for any $l,\ m \in \mathbb{N}$ and $f \in \mathcal{D}_K(\mathbb{R}^n)$, one has

$$p_{l,m}(f) = \max_{\substack{|\alpha|\le l \\ |\beta|\le m}} \sup_{x\in K}\left|x^\alpha D^\beta f(x)\right| \le \left(\max_{|\alpha|\le l}\sup_{x\in K}|x^\alpha|\right) p_{K,m}(f).$$

Whence the result.

(ii) We will use a truncating sequence. Let $\psi \in \mathcal{D}(\mathbb{R}^n)$ such that $0 \le \psi \le 1$ and $\psi = 1$ on $\overline{B}(0,1)$. Putting $\psi_j(x) = \psi\left(\dfrac{x}{j}\right)$ for every $j \in \mathbb{N}^*$, one obtains a sequence $(\psi_j)_j \subset \mathcal{D}(\mathbb{R}^n)$ such that $\psi_j = 1$ on $\overline{B}(0,j)$ and, for every $\beta \in \mathbb{N}^n$,

$$\sup_j \sup_{x\in\mathbb{R}^n}\left|D^\beta \psi_j(x)\right| \le \sup_{x\in\mathbb{R}^n}\left|D^\beta \psi(x)\right|.$$

Let now $f \in \mathcal{S}(\mathbb{R}^n)$. Put $f_j = f\psi_j$. One has $f_j \in \mathcal{D}(\mathbb{R}^n)$. Let us show that $(f_j)_j$ converges to f in $\mathcal{S}(\mathbb{R}^n)$. We have for any $\alpha, \beta \in \mathbb{N}^n$,

$$p_{\alpha,\beta}(f - f_j) = \sup_{x\in\mathbb{R}^n}\left|x^\alpha D^\beta (f - f_j)(x)\right| = \sup_{x\le j}\left|x^\alpha D^\beta\left[(1-\psi_j)f\right](x)\right|.$$

But

$$D^\beta(f - f_j) = \sum_{\gamma\le\beta} \frac{\beta!}{(\beta-\gamma)!\gamma!} D^{\beta-\gamma}(1-\psi_j) D^\gamma f.$$

Whence

$$p_{\alpha,\beta}(f - f_j) \le c_\beta \sum_{\gamma \le \beta} \frac{\beta!}{(\beta - \gamma)!\gamma!} \sup_{\|x\| \ge j} |x^\alpha D^\gamma f(x)|$$

where $c_\beta = \sum_{\gamma \le \beta} \sup_{x \in \mathbb{R}^n} \left| D^{\beta-\gamma}\psi(x) \right|$. Now, the second term tends to zero when j tends to $+\infty$.

$\square$

6.2   Multiplication operators

By the very definition, $\mathcal{S}(\mathbb{R}^n)$ is stable for the usual product and for multiplication by polynomials. It is not so for multiplication by functions of class $\mathcal{C}^\infty$. Indeed if $f(x) = e^{-x^2}$, one has $f \in \mathcal{S}(\mathbb{R})$ but $e^{x^2} e^{-x^2} = 1$ and the constant function, of value 1, does not belong to $\mathcal{S}(\mathbb{R})$. This leads us to the following notion.

Definition 6.1. An $f \in \mathcal{E}(\mathbb{R}^n)$ is said to be a multiplier for $\mathcal{S}(\mathbb{R}^n)$, if $f\varphi \in \mathcal{S}(\mathbb{R}^n)$ for every $\varphi \in \mathcal{S}(\mathbb{R}^n)$.

The set of multipliers for $\mathcal{S}(\mathbb{R}^n)$ will be denoted $\mathcal{O}_M(\mathbb{R}^n)$.

For $f \in \mathcal{E}(\mathbb{R}^n)$ to be a multiplier it is necessary and sufficient that for every $k \in \mathbb{N}$ and every $\beta \in \mathbb{N}^n$,

$$q_{k,\beta}(f\varphi) < +\infty, \text{ for every } \varphi \in \mathcal{S}(\mathbb{R}^n).$$

Let us look for a practical sufficient condition. One has

$$q_{k,\beta}(f\varphi) = \sup_{x \in \mathbb{R}^n} \left(1 + \|x\|^2\right)^k D^\beta(f\varphi)(x)$$

$$= \sup_{x \in \mathbb{R}^n} \left| \left(1 + \|x\|^2\right)^k \sum_{\gamma \le \beta} \frac{\beta!}{(\beta-\gamma)!\gamma!} D^\gamma f D^{\beta-\gamma}\varphi(x) \right|.$$

If there is $c > 0$ and $l \in \mathbb{N}$ such that

$$|D^\gamma f(x)| \le c \left(1 + \|x\|^2\right)^l; \ \gamma \le \beta \text{ and } x \in \mathbb{R}^n,$$

then

$$q_{k,\beta}(f\varphi) \le c \sup_{x \in \mathbb{R}^n} \left(1 + \|x\|^2\right)^{k+l} \sum_{\gamma \le \beta} \frac{\beta!}{(\beta-\gamma)!\gamma!} \left| D^{\beta-\gamma}\varphi(x) \right|$$

$$\le c' \max_{\gamma \le \beta} q_{k+l,\beta-\gamma}(\varphi).$$

So this raises the idea of the following definition.

Definition 6.2. A function $f : \mathbb{R} \longrightarrow \mathbb{C}$ is said to be slowly increasing on $\mathbb{R}$, if there is an $l \in \mathbb{N}$ such that the function

$$x \longmapsto \frac{f(x)}{\left(1 + \|x\|^2\right)^l}$$

is bounded. It is said to be of class $\mathcal{C}^\infty$ with slow increase if it is of class $\mathcal{C}^\infty$ and slowly increasing as well as its derivatives of any order.

We have the following result.

Proposition 6.9. *For* $f \in \mathcal{E}(\mathbb{R}^n)$, *the following assertions are equivalent.*

(i) f is of class $\mathcal{C}^\infty$ and slowly increasing.
(ii) $f \in \mathcal{O}_M(\mathbb{R}^n)$.

Proof. (i)$\Longrightarrow$(ii) By the calculations before Definition 6.2.
(ii)$\Longrightarrow$(i) Let $f \in \mathcal{O}_M(\mathbb{R}^n)$. If it is not of class $\mathcal{C}^\infty$ and slowly increasing, then one of its derivatives, say $D^\alpha f$, is not slowly increasing. As $D_g^\gamma \in \mathcal{O}_M(\mathbb{R}^n)$, for any γ, whenever $g \in \mathcal{O}_M(\mathbb{R}^n)$ it is sufficient to argue with f. We will show that if f is not slowly increasing, then there is $\varphi \in S(\mathbb{R}^n)$ such that $f\varphi \notin S(\mathbb{R}^n)$. The function

$$x \longmapsto \frac{f(x)}{\left(1 + \|x\|^2\right)^l}$$

is not bounded for any $l \in \mathbb{N}$. There is a sequence $(a_l)_{l \in \mathbb{N}}$ such that

$$\frac{f(a_l)}{\left(1 + \|a_l\|^2\right)^l} \text{ tends to } +\infty.$$

We look for φ such that $f\varphi$ does not tend to zero at infinity. As f is continuous (hence bounded on any compact set) one can choose $(a_l)_l$ such that $a_1, \ldots, a_{l-1} \in B(0, l)$, $a_l \notin B(0, l)$ and $\|a_l - a_{l-1}\| \geq 1$. It then suffices to find φ such that

$$(f\varphi)(a_l) = c \, \frac{f(a_l)}{\left(1 + \|a_l\|^2\right)^l}$$

for a given constant $c > 0$, i.e.,

$$\varphi(a_l) = \frac{c}{\left(1 + \|a_l\|^2\right)^l}.$$

But it is sufficient to take

$$\varphi(x) = \sum_{k \in \mathbb{N}} \frac{\theta(x - a_k)}{\left(1 + \|a_k\|^2\right)^k}, \text{ where } \theta \in \mathcal{D}\left(\mathbb{R}^n\right)$$

and such that $supp\ \theta = \overline{B}(0,1)$ (cf. Example 1.3 of Chapter 1). What remains to see is that $\varphi \in \mathcal{S}(\mathbb{R}^n)$. For this one has

$$x^\alpha D^\beta \varphi(x) = \sum_{k \in \mathbb{N}} \frac{x^\alpha D^\beta \theta(x - a_k)}{\left(1 + \|a_k\|^2\right)^k}$$

and

$$\sup_{x \in \mathbb{R}^n} \left|x^\alpha D^\beta \theta(x - a_k)\right| = \sup_{\|y\| \leq 1} (y + a_k)^\alpha D^\beta \theta(y)$$

$$\leq (1 + \|a_k\|)^{n|\alpha|} \sup_{\|y\| \leq 1} \left|D^\beta \theta(y)\right|$$

$$\leq c_1 (1 + \|a_k\|)^{n|\alpha|},$$

with $c_1 = \sup_{\|y\| \leq 1} \left|D^\beta \theta(y)\right|$. Hence

$$\sup \left|x^\alpha D^\beta \varphi(x)\right| \leq c_1 \sum_{k \in \mathbb{N}} \frac{(1 + \|a_k\|)^{n|\alpha|}}{\left(1 + \|a_k\|^2\right)^k} < +\infty.$$

$\square$

The following result is a consequence of the above.

Proposition 6.10. *For* $f \in \mathcal{O}_M(\mathbb{R}^n)$, *the linear map* $\varphi \longmapsto f\varphi$ *from* $\mathcal{S}(\mathbb{R}^n)$ *into* $\mathcal{S}(\mathbb{R}^n)$ *is continuous.*

We also have the following properties which are easily established.

Proposition 6.11.

(i) *The set* $\mathcal{O}_M(\mathbb{R}^n)$ *is a vector subspace of* $\mathcal{E}(\mathbb{R}^n)$. *It is also stable for the ordinary product.*

(ii) *Let* $f \in \mathcal{O}_M(\mathbb{R}^n)$. *Then for every* $a \in \mathbb{R}$ *and every* $\alpha \in \mathbb{N}^n$, *the functions*

$$\hat{f}, \tau_a f, D^\alpha f \text{ and } M^\alpha f$$

belong to $\mathcal{O}_M(\mathbb{R}^n)$.

6.3 Fourier transformation in $\mathcal{S}(\mathbb{R}^n)$

The space $\mathcal{S}(\mathbb{R}^n)$ has been constructed in order to be invariant by the Fourier transformation.

Proposition 6.12. *The Fourier transformation $\mathcal{F}$ is a continuous linear map of $\mathcal{S}(\mathbb{R}^n)$ into $\mathcal{S}(\mathbb{R}^n)$.*

Proof. For $f \in \mathcal{S}(\mathbb{R}^n)$, $\hat{f}$ is defined since $\mathcal{S}(\mathbb{R}^n) \subset L^1(\mathbb{R}^n)$. According to the exchange formulas (cf. Theorem 5.2 of Chapter 5), one has

$$(2i\pi M)^\alpha D^\beta \hat{f} = (2i\pi M)^\alpha \widehat{(-2i\pi M)^\beta} f = D^\alpha [\widehat{(-2i\pi M)^\beta f}],$$

and we have

$$p_{\alpha,\beta}(\hat{f}) \leq 2\pi^{|\beta|-|\alpha|} r_{\alpha,0}(M^\beta f).$$

So $\hat{f} \in \mathcal{S}(\mathbb{R}^n)$. The continuity of $\mathcal{F}$ follows from the previous inequality and the continuity of M^β (Proposition 6.5). $\qquad\square$

Remark 6.2. One shows, in the same manner, that the Fourier cotransformation $\overline{\mathcal{F}}$ is a linear continuous map of $\mathcal{S}(\mathbb{R}^n)$ into $\mathcal{S}(\mathbb{R}^n)$. We have even more.

Proposition 6.13. *The maps $\mathcal{F}$ and $\overline{\mathcal{F}}$ are topological isomorphisms of $\mathcal{S}(\mathbb{R}^n)$, reciprocal of each other, i.e., one has the following formulas, said to be of reciprocity,*

$$\mathcal{F}\overline{\mathcal{F}}f = \overline{\mathcal{F}}\mathcal{F}f = f, \textit{ for every } f \in \mathcal{S}(\mathbb{R}^n).$$

Proof. For $f \in \mathcal{S}(\mathbb{R}^n)$ one obviously has $f \in L^1(\mathbb{R}^n)$. And we have seen (cf. Proof of Proposition 5.2, Chapter 5) that for every $\varphi \in \mathcal{D}(\mathbb{R}^n)$, one has $\overline{\mathcal{F}}\mathcal{F}\varphi = \varphi$. Hence this equality takes place in $\mathcal{S}(\mathbb{R}^n)$, since $\overline{\mathcal{F}}$ and $\mathcal{F}$ are continuous linear mappings of $\mathcal{S}(\mathbb{R}^n)$ into $\mathcal{S}(\mathbb{R}^n)$, by Proposition 3.1, and the fact that $\mathcal{D}(\mathbb{R}^n)$ is dense in $\mathcal{S}(\mathbb{R}^n)$. $\qquad\square$

6.4 Tempered distributions

Since for every compact subset K of $\mathbb{R}^n$ the injection $\mathcal{D}_K(\mathbb{R}^n) \longrightarrow \mathcal{S}(\mathbb{R}^n)$ is continuous, the restriction of any continuous linear form on $\mathcal{S}(\mathbb{R}^n)$ to $\mathcal{D}(\mathbb{R}^n)$ is a distribution on $\mathbb{R}^n$.

Definition 6.3. Let T be a linear form on $\mathcal{D}(\mathbb{R}^n)$. We say that T is a tempered distribution on $\mathbb{R}^n$, if it is continuous for the topology induced by that one of $\mathcal{S}(\mathbb{R}^n)$.

Due to the density of $\mathcal{D}(\mathbb{R}^n)$ in $\mathcal{S}(\mathbb{R}^n)$, one has the following characterization.

Proposition 6.14. *Let $T : \mathcal{D}(\mathbb{R}^n) \longrightarrow \mathbb{C}$ be a linear form. The following assertions are equivalent.*

(i) T is a tempered distribution.
(ii) T extends to a continuous linear form on $\mathcal{S}(\mathbb{R}^n)$.

Remark 6.3. The previous proposition allows the identification of the space of tempered distributions with the topological dual $\mathcal{S}'(\mathbb{R}^n)$ of $\mathcal{S}(\mathbb{R}^n)$.

Examples 6.1

1) **Distributions with compact support.** Any distribution with compact support is tempered, for the injection $\mathcal{S}(\mathbb{R}^n) \subset \mathcal{E}(\mathbb{R}^n)$ is continuous (cf. (ii) of Proposition 6.7). Hence one has $\mathcal{E}'(\mathbb{R}^n) \subset \mathcal{S}'(\mathbb{R}^n)$.

2) **Slowly increasing measures.** A locally integrable (even continuous) function f does not always define a tempered distribution (cf. Example 6 hereafter). Writing for every $\varphi \in \mathcal{D}(\mathbb{R}^n)$ and $k \in \mathbb{N}^*$,

$$\langle T_f, \varphi \rangle = \int_{\mathbb{R}^n} \frac{f(x)}{\left(1 + \|x\|^2\right)^k} \left(1 + \|x\|^2\right)^k \varphi(x)\, dx$$

and taking into account that $q_{k,0}(\varphi) = \sup_{x \in \mathbb{R}^n} \left(1 + \|x\|^2\right)^k |\varphi(x)|$, one sees that a sufficient condition for T_f to be tempered is the existence of a k such that

$$\int_{\mathbb{R}^n} \frac{1}{\left(1 + \|x\|^2\right)^k} |f(x)|\, dx < +\infty.$$

In this case, the measure T_f is said to be slowly increasing. More generally, a Radon measure μ on $\mathbb{R}$ is said to be of slow increase if there is $k \in \mathbb{N}^*$ such that

$$c = \int_{\mathbb{R}^n} \frac{1}{\left(1 + \|x\|^2\right)^k} d\,|\mu| < +\infty.$$

Now, any measure of slow increase is a tempered distribution: Indeed

$$|\langle \mu, \varphi \rangle| = \left| \int_{\mathbb{R}^n} \varphi(x) d\mu(x) \right| \le c \, q_{k,0}(\varphi).$$

Moreover, its extension to $\mathcal{S}(\mathbb{R}^n)$ is exactly the linear form

$$f \longmapsto \int_{\mathbb{R}^n} f(x) d\mu.$$

First, every $f \in \mathcal{S}(\mathbb{R}^n)$ is μ-integrable, as the product of the bounded μ-measurable function

$$x \longmapsto f(x) \left(1 + \|x\|^2\right)^k$$

and the μ-integrable function

$$x \longmapsto \frac{1}{\left(1 + \|x\|^2\right)^k}.$$

On the other hand, the restriction of the linear form to $\mathcal{D}(\mathbb{R}^n)$ in question is equal to μ.

The following particular case is worth mentioning.

3) **Multipliers.** Any $f \in \mathcal{O}_M(\mathbb{R}^n)$ defines a tempered distribution. Indeed by Proposition 6.9, there is $l \in \mathbb{N}$ such that the function

$$x \longmapsto \frac{f(x)}{\left(1 + \|x\|^2\right)^l} \text{ is bounded.}$$

Hence the measure T_f is of slow increase, since

$$\int_{\mathbb{R}^n} \frac{|f(x)|}{\left(1 + \|x\|^2\right)^{l+n}} dx < +\infty.$$

4) **Injection of $L^p(\mathbb{R}^n)$ into $\mathcal{S}'(\mathbb{R}^n)$.** Any space $L^p(\mathbb{R}^n)$, $1 \le p \le +\infty$, is continuously embedded in $\mathcal{S}'(\mathbb{R}^n)$ endowed with the vague topology. Indeed every $f \in L^p(\mathbb{R}^n)$ defines a tempered distribution for, by Hölder inequality, one has

$$\int_{\mathbb{R}^n} \frac{f(x)}{\left(1 + \|x\|^2\right)^n} dx \le \|f\|_p \left(\int \frac{1}{\left(1 + \|x\|^2\right)^{nq}} dx \right)^{1/q} \le \pi^n \|f\|_p$$

where $\frac{1}{p} + \frac{1}{q} = 1$. Moreover the injection $L^p(\mathbb{R}^n) \longrightarrow \mathcal{S}'(\mathbb{R}^n)$ is continuous since, for every $\varphi \in \mathcal{S}(\mathbb{R}^n)$,

$$p_\varphi(T_f) = |\langle T_f, \varphi \rangle| = \int_{\mathbb{R}^n} |f(x)| \, |\varphi(x)| \, dx \leq \pi^n \, \|f\|_p \, q_{n,0}(\varphi).$$

We also have the following particular case.

5) Any polynomial defines a tempered distribution. More generally, any measurable function dominated by a polynomial defines also a tempered distribution.

6) **Non tempered regular distributions.** The distribution defined by the function $f : x \longmapsto e^x$ is not tempered. Indeed, if it were so there should exist $c > 0$ and $m \in \mathbb{N}$ such that

$$|\langle T_f, \varphi \rangle| = \left| \int_{\mathbb{R}} e^x \varphi(x) \, dx \right| \leq c \max_{\substack{k \leq m \\ l \leq m}} \sup_{x \in \mathbb{R}} \left| x^k \varphi^{(l)}(x) \right|.$$

But for every $t \in \mathbb{R}$,

$$\int_{\mathbb{R}} e^x \varphi(x) dx = e^t \int_{\mathbb{R}} e^u \varphi(u+t) du.$$

In particular for ψ_t with $\psi_t(u) = \varphi(u - t)$, one has

$$\langle T_f, \psi_t \rangle = e^t \int_{\mathbb{R}} e^u \varphi(u) \, du = c_\varphi e^t.$$

On the other hand

$$\left| x^k \psi_t^{(l)}(x) \right| = \left| x^k \varphi^{(l)}(x - t) \right| = \left| (u+t)^k \varphi^{(l)}(u) \right|.$$

Taking $\varphi \in \mathcal{D}(\mathbb{R})$ with support in $[0, 1]$, one obtains

$$c_\varphi e^t \leq c_1 (1 + t)^m, \text{ for every } t$$

where c_1 is a constant; and this can not be valid.

The following stability results allow the widening of the class of tempered distributions.

Proposition 6.15. *Let $T \in \mathcal{S}'(\mathbb{R}^n)$. Then*

(i) *For any $\alpha \in \mathbb{N}^n$, the distributions $M^\alpha T$ and $D^\alpha T$ are tempered.*

(ii) *The maps $T \longmapsto M^\alpha T$ and $T \longmapsto D^\alpha T$ are continuous from $\mathcal{S}'(\mathbb{R}^n)$ into $\mathcal{S}'(\mathbb{R}^n)$.*

(iii) If $S \in \mathcal{S}'(\mathbb{R}^l)$ and $T \in \mathcal{S}'(\mathbb{R}^n)$, then $S \otimes T \in \mathcal{S}'(\mathbb{R}^{l+n})$.

Proof.

(i) $M^\alpha T$ is continuous on $\mathcal{S}(\mathbb{R}^n)$ as the composite of the continuous maps M^α (cf. Proposition 6.5) and T. The same for the other.

(ii) Results from $p_\varphi(M^\alpha T) = p_{M^\alpha \varphi}(T)$ and $p_\varphi(D^\alpha T) = p_{D^\alpha \varphi}(T)$.

(iii) For every $\theta \in \mathcal{D}(\mathbb{R}^l \times \mathbb{R})$, put $\psi(y) = \langle S, \theta(., y) \rangle$. As $S \in \mathcal{S}'(\mathbb{R}^l)$, there is a constant $c > 0$ and $(k_0, m_0) \in \mathbb{N} \times \mathbb{N}$, such that

$$|\langle S, \varphi \rangle| \le c\, p_{k_0, m_0}(\varphi); \quad \varphi \in \mathcal{D}(\mathbb{R}^l).$$

Therefore, we obtain

$$p_{k,m}(\psi) \le c p_{k_0+k, m_0+m}(\theta).$$

This shows that the map $\theta \longmapsto \psi$ is continuous from $\mathcal{D}(\mathbb{R}^l \times \mathbb{R}^n)$ into $\mathcal{D}(\mathbb{R}^n)$ when $\mathcal{D}(\mathbb{R}^l \times \mathbb{R}^n)$ is endowed with the topology of $\mathcal{S}(\mathbb{R}^l \times \mathbb{R}^n)$ and $\mathcal{D}(\mathbb{R}^n)$ by that of $\mathcal{S}(\mathbb{R}^n)$. It follows that the linear form $\theta \longmapsto \langle T, \psi \rangle$ is continuous on $\mathcal{D}(\mathbb{R}^l \times \mathbb{R}^n)$ for the topology of $\mathcal{S}(\mathbb{R}^l \times \mathbb{R}^n)$, i.e., $S \otimes T \in \mathcal{S}'(\mathbb{R}^l \times \mathbb{R}^n)$.

$\square$

Concerning the product of a tempered distribution by a function, one has the following: If $f \in \mathcal{S}(\mathbb{R}^n)$ then for any $T \in \mathcal{S}'(\mathbb{R}^n)$ one evidently has $fT \in \mathcal{D}'(\mathbb{R}^n)$. Moreover for every $\varphi \in \mathcal{D}(\mathbb{R}^n)$, $\langle fT, \varphi \rangle = \langle T, f\varphi \rangle$. And the distribution fT is continuous on $\mathcal{D}(\mathbb{R}^n)$ for the topology of $\mathcal{S}(\mathbb{R}^n)$, as the composite of continuous mappings. In fact what went before is valid whenever $f \in \mathcal{O}_M(\mathbb{R}^n)$. Moreover the map $T \longmapsto fT$, of $\mathcal{S}'(\mathbb{R}^n)$ into $\mathcal{S}'(\mathbb{R}^n)$, is continuous due to $p_\varphi(fT) = p_{f\varphi}(T)$. So we have the following result.

Proposition 6.16. *Let* $f \in \mathcal{O}_M(\mathbb{R}^n)$. *Then* $fT \in \mathcal{S}'(\mathbb{R}^n)$, *for every* $T \in \mathcal{S}'(\mathbb{R}^n)$, *and the map* $T \longmapsto fT$ *of* $\mathcal{S}'(\mathbb{R}^n)$ *into* $\mathcal{S}'(\mathbb{R}^n)$ *is continuous.*

Remark 6.4. 6) of Examples 6.1 shows that a distribution of finite order is not necessarily tempered; however any tempered distribution T is of finite order. Indeed there are $l, m \in \mathbb{N}$ and $c > 0$ such that for every $\varphi \in \mathcal{D}(\mathbb{R}^n)$

$$|\langle T, \varphi \rangle| \le c\, q_{l,m}(\varphi)$$

where

$$q_{l,m}(\varphi) = \max_{|\alpha| \le m} \sup_{x \in \mathbb{R}^n} \left(1 + \|x\|^2\right)^l |D^\alpha \varphi(x)|.$$

Then for any compact K in $\mathbb{R}^n$ and every $\varphi \in \mathcal{D}_K(\mathbb{R}^n)$, one has

$$|\langle T, \varphi \rangle| \le c_1 p_{K,m}(\varphi) \text{ where } c_1 = c \sup_{x \in K} \left(1 + \|x\|^2\right)^l.$$

So T is of order less or equal to m.

The following diagram recapitulates the injections between spaces considered in the preceding.

$$\mathcal{E}'(\mathbb{R}^n)$$

$$\mathcal{D}(\mathbb{R}^n) \qquad\qquad\qquad\qquad \mathcal{S}'(\mathbb{R}^n).$$

$$\mathcal{S}(\mathbb{R}^n) \quad \longrightarrow \quad L^p(\mathbb{R}^n)$$

6.5   Convolution operators

We have seen that the convolution product acts on $\mathcal{S}(\mathbb{R}^n)$ (cf. Section 6.1, Proposition 6.4). On the contrary the convolution product of two distributions, even when one of them is a function of class $\mathcal{C}^\infty$, is not always defined. But we have seen (cf. Proposition 4.6 of Chapter 4) that if T is a distribution with compact support, then one can take the convolution of T with any function $f \in \mathcal{C}^\infty(\mathbb{R}^n)$ by $(T * f)(t) = \langle T_x, f(t-x) \rangle$. So we have for every $\varphi \in \mathcal{D}(\mathbb{R}^n)$,

$$\langle T * f, \varphi \rangle = \int_{\mathbb{R}^n} \langle T_x, f(t-x) \rangle \, \varphi(t) dt = \left\langle T_x, \int_{\mathbb{R}^n} f(t-x) \varphi(t) dt \right\rangle.$$

Thus, for every $\varphi \in \mathcal{D}(\mathbb{R}^n)$

$$\langle T * f, \varphi \rangle = \left\langle T, \overset{\vee}{f} * \varphi \right\rangle.$$

As a matter of fact, $\left\langle T, \overset{\vee}{f} * \varphi \right\rangle$ is meaningful whenever $f, \varphi \in \mathcal{S}(\mathbb{R}^n)$ and T is a tempered distribution on $\mathbb{R}$.

Definition 6.4. Let $T \in \mathcal{S}'(\mathbb{R}^n)$ and $f \in \mathcal{S}(\mathbb{R}^n)$. The convolution product of T by f is the tempered distribution, denoted by $T * f$ or $f * T$, given by

$$\langle T * f, \varphi \rangle = \left\langle T, \overset{\vee}{f} * \varphi \right\rangle; \ \varphi \in \mathcal{S}(\mathbb{R}^n).$$

The following properties are easily verified.

Proposition 6.17. *Let $T \in \mathcal{S}'(\mathbb{R}^n)$ and $f \in \mathcal{S}(\mathbb{R}^n)$. For any $a \in \mathbb{R}^n$ and any $\alpha \in \mathbb{N}^n$, one has*

(i) $\overset{\vee}{T} * \overset{\vee}{f} = (T * f)^{\vee}.$
(ii) $\tau_a(T * f) = (\tau_a T) * f.$
(iii) $D^\alpha(T * f) = (D^\alpha T) * f.$

Remark 6.5. The linear map $T \longmapsto T * f$ of $\mathcal{S}'(\mathbb{R}^n)$ into $\mathcal{S}'(\mathbb{R}^n)$ is continuous, for $p_\varphi(T * f) = p_{\overset{\vee}{f} * \varphi}(T).$

We can also consider the linear map $f \longmapsto T * f$ of $\mathcal{S}(\mathbb{R}^n)$ into $\mathcal{S}'(\mathbb{R}^n)$. Its range is not necessarily in $\mathcal{S}(\mathbb{R}^n)$. This leads us to the notion of a convolutor.

Definition 6.5. We say that a tempered distribution T on $\mathbb{R}$ is a convolution operator or a convolutor if, for every $f \in \mathcal{S}(\mathbb{R}^n)$, $f * T \in \mathcal{S}(\mathbb{R}^n)$ and if moreover the map $f \longmapsto f * T$ of $\mathcal{S}(\mathbb{R}^n)$ into $\mathcal{S}(\mathbb{R}^n)$ is continuous.

The set of convolutors for $\mathcal{S}(\mathbb{R}^n)$ is denoted by $\mathcal{O}'_c(\mathbb{R}^n)$.

The verification of the following properties is straightforward (use Proposition 6.17).

Proposition 6.18. *The set $\mathcal{O}'_c(\mathbb{R}^n)$ is a vector subspace of $\mathcal{S}'(\mathbb{R}^n)$. Moreover if $T \in \mathcal{O}'_c(\mathbb{R}^n)$ then, for every $a \in \mathbb{R}^n$ and $\alpha \in \mathbb{N}^n$, the distributions $\overset{\vee}{T}$, $\tau_a T$, $D^\alpha T$ and $M^\alpha T$ are in $\mathcal{O}'_c(\mathbb{R}^n)$.*

Now here are interesting examples of convolutors.

Proposition 6.19. *Any distribution T with compact support on $\mathbb{R}^n$ is a convolutor for $\mathcal{S}(\mathbb{R}^n)$.*

Proof. Let $f \in \mathcal{S}(\mathbb{R}^n)$. By Proposition 4.6 of Chapter 4, $T * f$ is a function of class $\mathcal{C}^\infty$ given by

$$(T * f)(x) = \langle T_y, f(x - y) \rangle = \left\langle \overset{\vee}{T}_y, f(x + y) \right\rangle.$$

As $\overset{\vee}{T}$ is continuous on $\mathcal{E}(\mathbb{R}^n)$, there is a compact set L in $\mathbb{R}^n$, an integer $l \in \mathbb{N}$ and a positive constant c such that

$$\left| \left\langle \overset{\vee}{T}, g \right\rangle \right| \le c \, p_{L,l}(g); \, g \in \mathcal{E}(\mathbb{R}^n).$$

Now let $\alpha \in \mathbb{N}^n$ and $k \in \mathbb{N}$. One has

$$\left(1 + \|x\|^2\right)^k |D^\alpha(T * f)(x)| \leq c \sup_{y \in L} \max_{|\beta| \leq l} \left(1 + \|x\|^2\right)^k \left|D_x^\alpha D_y^\beta f(x + y)\right|.$$

Hence

$$q_{k,\alpha}(T * f) \leq c \sup_{y \in L} \sup_{x \in \mathbb{R}^n} \max_{|\gamma| \leq |\alpha| + l} \left(1 + \|x\|^2\right)^k |D^\gamma f(x + y)|$$

$$\leq c \sup_{y \in L} \sup_{x \in \mathbb{R}^n} \max_{|\gamma| \leq |\alpha| + l} \frac{\left(1 + \|x\|^2\right)^k}{\left(1 + \|x + y\|^2\right)^k} \left(1 + \|x + y\|^2\right)^k |D^\gamma f(x + y)|$$

$$\leq \left[c \sup_{y \in L} \left(1 + \|y\|\right)^{2k}\right] \max_{|\sigma| \leq |\alpha| + l} q_{k,\gamma}(f),$$

since, for every $x \in \mathbb{R}^n$ and $y \in L$,

$$\frac{1 + \|x\|^2}{1 + \|x + y\|^2} \leq (1 + \|y\|)^2.$$

Indeed, putting $u + v = x$ and $y = -v$, this inequality is equivalent to

$$1 + \|u + v\|^2 \leq \left(1 + \|u\|^2\right)(1 + \|v\|)^2.$$

The latter is true for $v = 0$. For $v \neq 0$ it results from

$$2 \|u\| \leq 2 \|u\|^2 + 2 + \|u\|^2 \|v\|.$$

Now, if $f \in L^1(\mathbb{R}^n)$ we have seen that $T_f \in \mathcal{S}'(\mathbb{R}^n)$ (cf. 4) of Examples 6.1. By Definition 6.4 one has, for every $g \in \mathcal{S}(\mathbb{R}^n)$,

$$\langle T_f * g, \varphi \rangle = \left\langle T_f, \overset{\vee}{g} * \varphi \right\rangle, \text{ for every } \varphi \in \mathcal{S}(\mathbb{R}^n).$$

Applying the Fubini theorem, one finds that

$$\left\langle T_f, \overset{\vee}{g} * \varphi \right\rangle = \langle T_{g*f}, \varphi \rangle.$$

So $T_f * g = T_{g*f}$. On the other hand, we have seen that if $f, g \in \mathcal{S}(\mathbb{R}^n)$, then for any $l, \alpha \in \mathbb{N}^n$,

$$r_{l,\alpha}(g * f) \leq c_l \, r_{l,\alpha}(g) \, r_{l,0}(f).$$

If $g \in \mathcal{S}(\mathbb{R}^n)$ and $f \in L^1(\mathbb{R}^n)$, the inequality above is still valid. Moreover if $r_{l,0}(f)$ is finite, then $g * f \in \mathcal{S}(\mathbb{R}^n)$ and the linear map $g \longmapsto g * f$ of $\mathcal{S}(\mathbb{R}^n)$ into $\mathcal{S}(\mathbb{R}^n)$ is continuous. So appears a sufficient condition on T_f to be a convolutor for $\mathcal{S}(\mathbb{R}^n)$.

$\square$

Definition 6.6. A function $f : \mathbb{R}^n \longrightarrow \mathbb{C}$ is said to be rapidly decreasing if $r_{\alpha,0}(f) < +\infty$, for $\alpha \in \mathbb{N}^n$, i.e., $\displaystyle\int_{\mathbb{R}^n} |x^\alpha f(x)| \, dx < +\infty$.

Remark 6.6. Taking $\alpha = (0, ..., 0)$, one sees that any function with rapid decrease is necessarily in $L^1(\mathbb{R}^n)$.

Proposition 6.20. *Any regular distribution defined by a function with rapid decrease is a convolutor for* $\mathcal{S}(\mathbb{R}^n)$.

We will see that the elements of $\mathcal{O}'_c(\mathbb{R}^n)$ can be considered as convolutors for $\mathcal{S}'(\mathbb{R}^n)$. If $T \in \mathcal{O}'_c(\mathbb{R}^n)$, one has also $\overset{\vee}{T} \in \mathcal{O}'_c(\mathbb{R}^n)$. Hence $\overset{\vee}{T} * f \in \mathcal{S}(\mathbb{R}^n)$, for every $f \in \mathcal{S}(\mathbb{R}^n)$. Then for any $S \in \mathcal{S}'(\mathbb{R}^n)$, $\left\langle S, \overset{\vee}{T} * f \right\rangle$ has always a meaning.

Definition 6.7. Let $T \in \mathcal{O}'_c(\mathbb{R}^n)$ and $S \in \mathcal{S}'(\mathbb{R}^n)$. The convolution product of T by S is the tempered distribution, denoted $T * S$, given by

$$\langle T * S, f \rangle = \left\langle S, \overset{\vee}{T} * f \right\rangle, \quad f \in \mathcal{S}(\mathbb{R}^n).$$

Remark 6.7.

1) The bilinear mapping $(T, S) \longmapsto T * S$, of $\mathcal{O}'_c(\mathbb{R}^n) \times \mathcal{S}'(\mathbb{R}^n)$ into $\mathcal{S}'(\mathbb{R}^n)$, is separately continuous since for every $\varphi \in \mathcal{S}(\mathbb{R}^n)$,

$$p_\varphi(T * S) = p_{\overset{\vee}{T} * \varphi}(S) = p_{\overset{\vee}{S} * \varphi}(T).$$

2) For any $T \in \mathcal{O}'_c(\mathbb{R}^n)$, the linear map $S \longmapsto S * T$ of $\mathcal{S}'(\mathbb{R}^n)$ into $\mathcal{S}'(\mathbb{R}^n)$ is continuous. We say that T is a convolutor for $\mathcal{S}'(\mathbb{R}^n)$.

The convolution product of two tempered distributions is not always defined. However one has the following result.

Proposition 6.21. *The set* $\mathcal{D}'(\mathbb{R}_+) \cap \mathcal{S}'(\mathbb{R})$ *is a subalgebra of the convolution algebra* $\mathcal{D}'(\mathbb{R}_+)$.

Proof. It remains to show that $S * T \in \mathcal{S}'(\mathbb{R})$ for any $S, T \in \mathcal{D}'(\mathbb{R}_+) \cap \mathcal{S}'(\mathbb{R})$. Taking into account the relation

$$S * T = \tau_{-2a}(\tau_a S * \tau_a T) \text{ for } a \in \mathbb{R},$$

we can suppose that *supp* S and *supp* T are included in $[a, +\infty[$ where $a > 0$. Now we show that the linear form

$$\varphi \longmapsto \langle S * T, \varphi \rangle$$

is continuous on $\mathcal{D}(\mathbb{R})$ for the topology of $\mathcal{S}(\mathbb{R})$. Let $\varphi \in \mathcal{D}(\mathbb{R})$. As *supp* S and *supp* T are included in $[a, +\infty[$ with $a > 0$, a separation lemma of Urysohn type (cf. [15]) provides a function $\chi \in \mathcal{D}(\mathbb{R} \times \mathbb{R})$ such that $0 \leq \chi \leq 1$, *supp* $\chi \subset \mathbb{R}_+ \times \mathbb{R}_+$ and $\chi = 1$ on a neighborhood of the compact set $(supp\, S \times supp\, T) \cap supp\, \varphi^\Delta$. By the very definition of $*$, one has

$$\langle S * T, \varphi \rangle = \langle S \otimes T, \chi\varphi^\Delta \rangle.$$

Next $S \otimes T \in \mathcal{S}'(\mathbb{R}^2)$ by (iii) of Proposition 6.15. Moreover

$$\sup_{(x,y)\in\mathbb{R}^2} \left|(1+x^2+y^2)^k \varphi(x+y)\chi(x,y)\right| \leq \sup_{(x,y)\in\mathbb{R}_+^2} \left|(1+x^2+y^2)^k \varphi(x+y)\right|$$

$$\leq \sup_{(x,y)\in\mathbb{R}_+^2} \left|\left(1+|x+y|^2\right)^k \varphi(x+y)\right|$$

$$\leq q_{k,0}(\varphi).$$

Whence the result. $\qquad\square$

6.6   Fourier transformation of tempered distributions

A justification of the following definition has been given at the beginning of Section 6.1.

Definition 6.8. For any $T \in \mathcal{S}'(\mathbb{R}^n)$, put

$$\langle \mathcal{F}T, \varphi \rangle = \langle T, \mathcal{F}\varphi \rangle, \quad \varphi \in \mathcal{S}(\mathbb{R}^n)$$

and

$$\langle \overline{\mathcal{F}}T, \varphi \rangle = \langle T, \overline{\mathcal{F}}\varphi \rangle, \quad \varphi \in \mathcal{S}(\mathbb{R}^n).$$

We say that the distribution $\mathcal{F}T$, also denoted by $\hat{T}$, is the Fourier transform of T. The distribution $\overline{\mathcal{F}}T$ is called the Fourier cotransform of T.

The link between Fourier transformations in $\mathcal{S}'(\mathbb{R}^n)$ and its subspace $L^1(\mathbb{R}^n)$ is given as follows.

Proposition 6.22. *Fourier transformation in* $\mathcal{S}'(\mathbb{R}^n)$ *extends the one in* $L^1(\mathbb{R}^n)$, *i.e.,* $\hat{T}_f = T_{\hat{f}}$, *for every* $f \in L^1(\mathbb{R}^n)$.

Proof. Let $f \in L^1(\mathbb{R}^n)$. For $\varphi \in \mathcal{D}(\mathbb{R}^n)$, one has

$$\langle \mathcal{F}T_f, \varphi \rangle = \langle T_f, \mathcal{F}\varphi \rangle = \int_{\mathbb{R}^n} f(x) \mathcal{F}\varphi(x) dx$$

and

$$\left\langle T_{\hat{f}}, \varphi \right\rangle = \int_{\mathbb{R}^n} \hat{f}(x)\varphi(x) dx.$$

Conclude using the transfer formula.

$\square$

We will now see that if T is a distribution with compact support, then $\mathcal{F}T$ is in fact a function that coincides with the Fourier transform of T, seen as a function (cf. the beginning of Section 6.1).

Proposition 6.23. *Let $T \in \mathcal{E}'(\mathbb{R}^n)$ be a distribution with compact support. Then $\hat{T}$ is a function of class $\mathcal{C}^\infty$, given by*

$$\hat{T}(y) = \left\langle T_x, \ e^{-2i\pi xy} \right\rangle,$$

such that

$$D^\beta(\hat{T}) = \widehat{(-2i\pi M)^\beta T}.$$

Proof. For $\varphi \in \mathcal{D}(\mathbb{R}^n)$, one has

$$\begin{aligned}
\left\langle \hat{T}, \varphi \right\rangle &= \left\langle T, \overset{\wedge}{\varphi} \right\rangle \\
&= \left\langle T_x, \left\langle \varphi_y, e^{-2i\pi xy} \right\rangle \right\rangle \\
&= \left\langle \varphi_y, \left\langle T_x, e^{-2i\pi xy} \right\rangle \right\rangle \\
&= \left\langle \left\langle T_x, e^{-2i\pi xy} \right\rangle, \varphi_y \right\rangle \\
&= \left\langle T_g, \varphi \right\rangle
\end{aligned}$$

where $g(y) = \left\langle T_x, e^{-2i\pi xy} \right\rangle$. Hence $\hat{T} = T_g$.

The second assertion follows from (ii) of Corollary 3.1 of Chapter 3.

$\square$

Using Proposition 6.13, one obtains the following:

Proposition 6.24. *For any $T \in \mathcal{S}'(\mathbb{R}^n)$,*

$$(\mathcal{F}\overline{\mathcal{F}})T = (\overline{\mathcal{F}}\mathcal{F})T = T.$$

Remark 6.8. Putting $\left\langle \overset{\vee}{T}, \varphi \right\rangle = \left\langle T, \overset{\vee}{\varphi} \right\rangle$, one easily verifies that $\overset{\vee}{\mathcal{F}T} = \overline{\mathcal{F}}T$ and $\overline{\mathcal{F}}\overset{\vee}{T} = \mathcal{F}T$.

The reciprocity formula can then also be written $\mathcal{F}\mathcal{F}T = \overset{\vee}{T}$.

The following formulas are direct application of Proposition 6.23 and Remark 6.8, where $\chi_a(x) = e^{2\pi i a x}$.

(1) $\mathcal{F}\delta = 1$, $\mathcal{F}(\delta_a) = \overline{\chi_a}$ and $\mathcal{F}(D^\beta \delta) = (2i\pi M)^\beta$.
(2) $\mathcal{F}1 = \delta$, $\mathcal{F}(\chi_a) = \delta_a$ and $\mathcal{F}\left((-2i\pi M)^\beta\right) = D^\beta \delta$.

Here is a generalization of the Exchange Theorem ((vii) of Proposition 5.1 in Chapter 5). We begin with an important particular case.

Proposition 6.25. *Let* $\varphi \in \mathcal{S}(\mathbb{R}^n)$ *and* $T \in \mathcal{S}'(\mathbb{R}^n)$. *Then*

(i) $\mathcal{F}(\varphi * T) = \mathcal{F}\varphi \mathcal{F}T$.
(ii) $\mathcal{F}(\varphi T) = \mathcal{F}\varphi * \mathcal{F}T$.

Proof.

(i) According to Definition 6.4, $\varphi * T \in \mathcal{S}'(\mathbb{R}^n)$. And for every $\psi \in \mathcal{S}(\mathbb{R}^n)$,

$$
\begin{aligned}
\langle \mathcal{F}(\varphi * T), \psi \rangle &= \langle \varphi * T, \mathcal{F}\psi \rangle \\
&= \left\langle T, \overset{\vee}{\varphi} * \mathcal{F}\psi \right\rangle \\
&= \left\langle \overline{\mathcal{F}}\mathcal{F}T, \overset{\vee}{\varphi} * \mathcal{F}\psi \right\rangle \\
&= \left\langle \mathcal{F}T, \overline{\mathcal{F}}(\overset{\vee}{\varphi} * \mathcal{F}\psi) \right\rangle \\
&= \left\langle \mathcal{F}T, \overline{\mathcal{F}}\overset{\wedge}{\varphi}\overline{\mathcal{F}}\mathcal{F}\psi \right\rangle \\
&= \left\langle \mathcal{F}T, (\overline{\mathcal{F}}\overset{\wedge}{\varphi})\psi \right\rangle \\
&= \langle \mathcal{F}T, (\mathcal{F}\varphi)\psi \rangle \\
&= \langle \mathcal{F}\varphi \mathcal{F}T, \psi \rangle.
\end{aligned}
$$

(ii) Same as before.

$\square$

According to 3) of Examples 6.1, $T_f \in \mathcal{S}'(\mathbb{R}^n)$ for any $f \in \mathcal{O}_M(\mathbb{R}^n)$. Hence we can consider the Fourier transforms of these distributions. On the other hand, if $T \in \mathcal{O}'_c(\mathbb{R}^n)$ then, for every $\varphi \in \mathcal{S}(\mathbb{R}^n)$, the distribution $T * \varphi$ is regular and is defined by a function in $\mathcal{S}(\mathbb{R}^n)$. More precisely, according to Proposition 6.25, $\mathcal{F}T$ is a regular distribution defined by a function of class $\mathcal{C}^\infty$. Actually the result is the following.

Proposition 6.26 (Exchange Theorem). *Let* $f \in \mathcal{O}_M(\mathbb{R}^n)$, $T \in \mathcal{O}'_c(\mathbb{R}^n)$ *and* $S \in \mathcal{S}'(\mathbb{R}^n)$. *Then*

(i) $\mathcal{F}T = T_g$, *with* $g \in \mathcal{O}_M(\mathbb{R}^n)$ *and* $\mathcal{F}T_f \in \mathcal{O}'_c(\mathbb{R}^n)$.
(ii) $\mathcal{F}(S * T) = \mathcal{F}T\mathcal{F}S$ *and* $\mathcal{F}(fS) = \mathcal{F}T_f * \mathcal{F}S$.

Proof.

(i) Since $T \in \mathcal{O}'_c(\mathbb{R}^n)$, we know that $\mathcal{F}T = T_g$ with $g \in \mathcal{C}^\infty(\mathbb{R}^n)$. It remains to show that g is a multiplier for $\mathcal{S}(\mathbb{R}^n)$, i.e., $g\varphi \in \mathcal{S}(\mathbb{R}^n)$, for every $\varphi \in \mathcal{S}(\mathbb{R}^n)$. According to the previous proposition, one has

$$\varphi(\mathcal{F}T) = \mathcal{F}\left[(\overline{\mathcal{F}}\varphi) * T\right].$$

Now

$$\varphi(\mathcal{F}T) = \varphi T_g = T_{g\varphi}.$$

And

$$\mathcal{F}\left[(\overline{\mathcal{F}}\varphi) * T\right] = \mathcal{F}T_h = T_{\mathcal{F}h} \text{ with } h \in \mathcal{S}(\mathbb{R}^n).$$

Whence

$$g\varphi = \mathcal{F}h \in \mathcal{S}(\mathbb{R}^n).$$

In an analoguous manner we show that $\mathcal{F}T_f \in \mathcal{O}'_c(\mathbb{R}^n)$ from the equality

$$\varphi * \mathcal{F}T_f = \mathcal{F}\left[(\overline{\mathcal{F}}\varphi) * T_f\right].$$

(ii) By Definition 6.7, $S * T \in \mathcal{S}'(\mathbb{R}^n)$, and according to Proposition 6.16, $fS \in \mathcal{S}'(\mathbb{R}^n)$. We proceed then as for the assertions of Proposition 6.25. $\qquad\square$

Remark 6.9. The two previous propositions are also valid for the Fourier cotransformation $\overline{\mathcal{F}}$.

This section ends with a consequence of Proposition 6.26.

Corollary 6.1. *For $T \in \mathcal{S}'(\mathbb{R}^n)$, the following assertions are equivalent.*

(1) $T \in \mathcal{O}'_c(\mathbb{R}^n)$.
(2) $T = \mathcal{F}T_f$ with $f \in \mathcal{O}_M(\mathbb{R}^n)$.

6.7 Ultradistributions

For $T \in \mathcal{S}'(\mathbb{R}^n)$, one has

$$\langle \mathcal{F}T, \varphi \rangle = \langle T, \mathcal{F}\varphi \rangle \, ; \; \varphi \in \mathcal{S}(\mathbb{R}^n).$$

The right-hand side has a meaning for an arbitrary T in $\mathcal{D}'(\mathbb{R}^n)$, whenever $\mathcal{F}\varphi \in \mathcal{D}(\mathbb{R}^n)$. We have seen, in Section 6.1, that if $\varphi \in \mathcal{D}(\mathbb{R}^n)$ then $\mathcal{F}\varphi \notin \mathcal{D}(\mathbb{R}^n)$ except $\varphi \equiv 0$. But we may have $\mathcal{F}\varphi \in \mathcal{D}(\mathbb{R}^n)$ for $\varphi \in \mathcal{S}(\mathbb{R}^n)$. We say that such a function is slowly decreasing and with compact support. Put

$$\mathcal{Z}(\mathbb{R}^n) = \{\varphi \in \mathcal{S}(\mathbb{R}^n) : \mathcal{F}\varphi \in \mathcal{D}(\mathbb{R}^n)\}.$$

This vector space is stable with respect to derivation and monomial multiplication (cf. Proposition 6.5). Notice also that

$$\mathcal{Z}(\mathbb{R}^n) = \{\varphi \in \mathcal{S}(\mathbb{R}^n) : \overline{\mathcal{F}}\varphi \in \mathcal{D}(\mathbb{R}^n)\}.$$

The map $\mathcal{F} : \mathcal{Z}(\mathbb{R}^n) \longrightarrow \mathcal{D}(\mathbb{R}^n)$ is a bijection. Indeed, it is one-to-one, for $\mathcal{F}$ is so on $\mathcal{S}(\mathbb{R}^n)$. It is onto, for $\mathcal{F}(\overline{\mathcal{F}}\varphi) = \varphi$ for every $\varphi \in \mathcal{D}(\mathbb{R}^n)$.

Definition 6.9. The topology of $\mathcal{Z}(\mathbb{R}^n)$ is the inverse image by $\mathcal{F}$ of the topology of $\mathcal{D}(\mathbb{R}^n)$.

The vector space $\mathcal{Z}(\mathbb{R}^n)$ is thus endowed with a topology of a separated *l.c.s.*. As the topology of $\mathcal{D}(\mathbb{R}^n)$ is stronger than that of $\mathcal{S}(\mathbb{R}^n)$, the topology of $\mathcal{Z}(\mathbb{R}^n)$ is then stronger than that induced by that one of $\mathcal{S}(\mathbb{R}^n)$.

Proposition 6.27. *The space $\mathcal{Z}(\mathbb{R}^n)$, endowed with its topology, is dense in $\mathcal{S}(\mathbb{R}^n)$.*

Proof.   This follows from the fact that $\mathcal{Z}(\mathbb{R}^n)$ and $\mathcal{D}(\mathbb{R}^n)$ are topologically isomorphic and the density of $\mathcal{D}(\mathbb{R}^n)$ in $\mathcal{S}(\mathbb{R}^n)$ (cf. (ii) of Proposition 6.8). $\qquad\square$

Definition 6.10. An ultradistribution is a continuous linear form on $\mathcal{Z}(\mathbb{R}^n)$. The space of ultradistributions is denoted $\mathcal{Z}'(\mathbb{R}^n)$.

Remark 6.10. One has $\mathcal{S}'(\mathbb{R}^n) \subset \mathcal{Z}'(\mathbb{R}^n)$ since $\mathcal{Z}(\mathbb{R}^n)$ is dense in $\mathcal{S}(\mathbb{R}^n)$.

The above allows a definition of the Fourier transform of an arbitrary distribution. The map $\mathcal{F} : \mathcal{Z}(\mathbb{R}^n) \longrightarrow \mathcal{D}(\mathbb{R}^n)$ is a topological isomorphism. Let $T \in \mathcal{D}'(\mathbb{R}^n)$. For every $\varphi \in \mathcal{Z}(\mathbb{R}^n)$, one has $\mathcal{F}\varphi \in \mathcal{D}(\mathbb{R}^n)$ and hence $\langle T, \mathcal{F}\varphi \rangle$ is meaningful.

Definition 6.11. For any $T \in \mathcal{D}'(\mathbb{R}^n)$, the Fourier transform of T, denoted by $\mathcal{F}T$, is the element of $\mathcal{Z}'(\mathbb{R}^n)$ given by

$$\langle \mathcal{F}T, \varphi \rangle = \langle T, \mathcal{F}\varphi \rangle, \ \varphi \in \mathcal{Z}(\mathbb{R}^n).$$

Remark 6.11. The Fourier transformation on $\mathcal{D}'(\mathbb{R}^n)$ is the transpose of the isomorphism $\mathcal{F} : \mathcal{Z}(\mathbb{R}^n) \longrightarrow \mathcal{D}(\mathbb{R}^n)$.

Remark 6.12. Using the isomorphism $\mathcal{F} : \mathcal{D}(\mathbb{R}^n) \longrightarrow \mathcal{Z}(\mathbb{R}^n)$, one defines, in the same way, the Fourier transformation on $\mathcal{Z}'(\mathbb{R}^n)$ by

$$\langle \mathcal{F}T, \varphi \rangle = \langle T, \mathcal{F}\varphi \rangle, \varphi \in \mathcal{D}(\mathbb{R}^n).$$

It is the transpose of the isomorphism $\mathcal{F} : \mathcal{D}(\mathbb{R}^n) \longrightarrow \mathcal{Z}(\mathbb{R}^n)$.

Remark 6.13. The Fourier transformation on $\mathcal{Z}'(\mathbb{R}^n)$ extends the one on $\mathcal{S}'(\mathbb{R}^n)$.

Remark 6.14. Considering the isomorphisms $\mathcal{F} : \mathcal{Z}(\mathbb{R}^n) \longrightarrow \mathcal{D}(\mathbb{R}^n)$ and $\overline{\mathcal{F}} : \mathcal{D}(\mathbb{R}^n) \longrightarrow \mathcal{Z}(\mathbb{R}^n)$, one defines, in an analoguous manner, the Fourier cotransformations.

In order to establish essential properties of the Fourier transformation, we introduce appropriate spaces.

Let $T \in \mathcal{Z}'(\mathbb{R}^n)$ be such that $\mathcal{F}T \in \mathcal{E}'(\mathbb{R}^n)$. Using the fact that $\mathcal{F}$ and $\overline{\mathcal{F}}$ are reciprocal isomorphisms of each other, one obtains $T = T_f$ with f of class $\mathcal{C}^\infty$ (cf. Proposition 6.23). This leads us to introduce the following space

$$\mathcal{O}(\mathbb{R}^n) = \{f \in \mathcal{E}(\mathbb{R}^n) : T_f \in \mathcal{Z}'(\mathbb{R}^n) \text{ and } \mathcal{F}T_f \in \mathcal{E}'(\mathbb{R}^n)\}.$$

A function f in $\mathcal{O}(\mathbb{R}^n)$ is said to have compact spectrum. We know, by Proposition 6.6, that $\mathcal{S}(\mathbb{R}^n) \subset L^p(\mathbb{R}^n)$ for every $p \geq 1$ and that any $L^p(\mathbb{R}^n)$

is embedded in $\mathcal{S}'(\mathbb{R}^n)$ by 4) of Examples 6.1. Moreover, $\mathcal{F}T_f \in \mathcal{E}'(\mathbb{R}^n)$ for every $f \in \mathcal{Z}(\mathbb{R}^n)$, by Proposition 6.22. Hence, one has

$$\mathcal{Z}(\mathbb{R}^n) \subset \mathcal{O}(\mathbb{R}^n) \subset \mathcal{E}(\mathbb{R}^n).$$

Now put

$$\mathcal{O}_1(\mathbb{R}^n) = \{T_f : f \in \mathcal{O}(\mathbb{R}^n)\}.$$

The spaces $\mathcal{O}_1(\mathbb{R}^n)$ and $\mathcal{O}(\mathbb{R}^n)$ are algebraically isomorphic by $b : T_f \longmapsto f$. The map $\mathcal{F} \circ b^{-1} : \mathcal{O}(\mathbb{R}^n) \longrightarrow \mathcal{E}'(\mathbb{R}^n)$ is an algebraic isomorphism. It is known that $\mathcal{E}(\mathbb{R}^n)$ is reflexive when $\mathcal{E}'(\mathbb{R}^n)$ is endowed with the strong dual topology, i.e., the topology of uniform convergence on the bounded subsets of $\mathcal{E}(\mathbb{R}^n)$. The latter is given by the family $(p_B)_B$ of seminorms, where B runs over the bounded subsets of $\mathcal{E}(\mathbb{R}^n)$, with

$$p_B(T) = \sup_{\varphi \in B} |\langle T, \varphi \rangle|.$$

Definition 6.12. The topology of $\mathcal{O}(\mathbb{R}^n)$ is the inverse image by $\mathcal{F} \circ b^{-1}$ of the strong dual topology of $\mathcal{E}'(\mathbb{R}^n)$.

Remark 6.15. The dual $\mathcal{O}'(\mathbb{R}^n)$ of $\mathcal{O}(\mathbb{R}^n)$ is algebraically isomorphic to $\mathcal{E}(\mathbb{R}^n)$. We endow it with the topology induced by that of $\mathcal{E}(\mathbb{R}^n)$.

Remark 6.16. One has $\mathcal{O}'(\mathbb{R}^n) \subset \mathcal{Z}'(\mathbb{R}^n)$, for the injection $\mathcal{Z}(\mathbb{R}^n) \longrightarrow \mathcal{O}(\mathbb{R}^n)$ is continuous.

Now we are going to look differently at the space $\mathcal{O}'(\mathbb{R}^n)$. For $f \in \mathcal{E}(\mathbb{R}^n)$, one has $\mathcal{F}(\overline{\mathcal{F}}T_f) = T_f$. Hence $T_f = \mathcal{F}(S)$ with $S \in \mathcal{Z}'(\mathbb{R}^n)$. We are thus led to consider the space

$$'\mathcal{O}(\mathbb{R}^n) = \{S \in \mathcal{Z}'(\mathbb{R}^n) : \mathcal{F}S \in \mathcal{E}_1(\mathbb{R}^n)\}$$

where $\mathcal{E}_1(\mathbb{R}^n) = \{T_f : f \in \mathcal{E}(\mathbb{R}^n)\}$. An element of $'\mathcal{O}(\mathbb{R}^n)$ is an ultradistribution with rapid decrease.

The Fourier transformation is an isomorphism of $'\mathcal{O}(\mathbb{R}^n)$ onto $\mathcal{E}_1(\mathbb{R}^n)$.

Remark 6.17. The dual $\mathcal{O}'(\mathbb{R}^n)$ of $\mathcal{O}(\mathbb{R}^n)$ is algebraically isomorphic to $'\mathcal{O}(\mathbb{R}^n)$.

Remark 6.18.

1) We also have the following injections

$$\mathcal{E}'(\mathbb{R}^n) \subset \mathcal{O}'(\mathbb{R}^n) \subset \mathcal{Z}'(\mathbb{R}^n).$$

2) We have the following equality

$$\mathcal{Z}_1\left(\mathbb{R}^n\right) = \mathcal{O}_1\left(\mathbb{R}^n\right) \cap {}'\mathcal{O}\left(\mathbb{R}^n\right).$$

Indeed, for $f \in \mathcal{Z}\left(\mathbb{R}^n\right)$ one has $T_f \in \mathcal{Z}'\left(\mathbb{R}^n\right)$ and $\mathcal{F}T_f = T_{\mathcal{F}_f}$. For the other inclusion, let $f \in \mathcal{O}\left(\mathbb{R}^n\right)$ with $\mathcal{F}T_f \in \mathcal{E}_1\left(\mathbb{R}^n\right)$. There is $g \in \mathcal{E}\left(\mathbb{R}^n\right)$ such that $\mathcal{F}T_f = T_g$. As $f \in \mathcal{O}\left(\mathbb{R}^n\right)$, one has $g \in \mathcal{D}\left(\mathbb{R}^n\right)$. Then $T_f = \overline{\mathcal{F}}\mathcal{F}T_f = T_{\overline{\mathcal{F}}g}$. Whence $\mathcal{F}f \in \mathcal{D}\left(\mathbb{R}^n\right)$ and hence $T_f \in \mathcal{Z}_1\left(\mathbb{R}^n\right)$.

Now we will see how $\mathcal{O}'(\mathbb{R}^n)$ acts on $\mathcal{O}(\mathbb{R}^n)$. For every $f \in \mathcal{O}\left(\mathbb{R}^n\right)$ and $T \in \mathcal{O}'\left(\mathbb{R}^n\right)$, one has

$$
\begin{aligned}
T(f) &= T\left(b(T_f)\right) \\
&= {}^t b(T(T_f)) \\
&= \left\langle T_f, {}^t b(T)\right\rangle; \quad \text{since } T_f \in \mathcal{O}_1\left(\mathbb{R}^n\right) \text{ and } {}^t b(T) \in \mathcal{O}'_1\left(\mathbb{R}^n\right) \\
&= \left\langle T_f, \mathcal{F}\overline{\mathcal{F}} \, {}^t b(T)\right\rangle \\
&= \left\langle T_f, \mathcal{F} \, {}^t(\mathcal{F}^{-1}) \, {}^t b(T)\right\rangle \\
&= \left\langle T_f, \mathcal{F}^t(b \circ \mathcal{F}^{-1})(T)\right\rangle \\
&= \left\langle \mathcal{F}T_f, {}^t(b \circ \mathcal{F}^{-1})(T)\right\rangle \\
&= \left\langle \mathcal{F}T_f, \Phi^{-1}(T)\right\rangle \qquad \text{where } \Phi = {}^t\left(\mathcal{F} \circ b^{-1}\right).
\end{aligned}
$$

Notice that we have $\mathcal{F}T_f \in \mathcal{E}_1\left(\mathbb{R}^n\right)$ and $\Phi^{-1}(T) \in \mathcal{E}\left(\mathbb{R}^n\right)$. Then

$$\left\langle \mathcal{F}T_f, \Phi^{-1}(T)\right\rangle = \langle T, f\rangle, \quad \text{i.e.,} \quad \left\langle \mathcal{F}T_f, \overline{\mathcal{F}} \circ {}^t b(T)\right\rangle = \langle T, f\rangle.$$

By abuse of notation, one writes

$$\left\langle \mathcal{F}T_f, \overline{\mathcal{F}} \circ {}^t b(T)\right\rangle = \left\langle \mathcal{F}T_f, \overline{\mathcal{F}}T\right\rangle.$$

It is clear that the product of two elements of $\mathcal{Z}\left(\mathbb{R}^n\right)$ is also in $\mathcal{Z}\left(\mathbb{R}^n\right)$. In fact we will see that $\mathcal{Z}\left(\mathbb{R}^n\right)$ is a module over $\mathcal{O}\left(\mathbb{R}^n\right)$.

Proposition 6.28.

1) $\mathcal{O}\left(\mathbb{R}^n\right) \subset \mathcal{O}_M\left(\mathbb{R}^n\right)$.

2) For every $f \in \mathcal{O}\left(\mathbb{R}^n\right)$, the map $\varphi \longmapsto f\varphi$, of $\mathcal{Z}\left(\mathbb{R}^n\right)$ into $\mathcal{Z}\left(\mathbb{R}^n\right)$, is continuous.

Proof.

1) Let $f \in \mathcal{O}\left(\mathbb{R}^n\right)$. According to Proposition 6.19, one has $\mathcal{F}T_f \in \mathcal{O}'_c\left(\mathbb{R}^n\right)$, i.e., $\mathcal{F}T_f$ is a convolutor. Hence $\mathcal{F}T_f = \mathcal{F}T_g$ with $g \in \mathcal{O}_M\left(\mathbb{R}^n\right)$, by Corollary 6.1. Whence the conclusion, for $\mathcal{F}$ is one-to-one on $\mathcal{Z}'\left(\mathbb{R}^n\right)$ and $f, g \in \mathcal{E}\left(\mathbb{R}^n\right)$.

2) According to 1), one has $f\varphi \in \mathcal{S}(\mathbb{R}^n)$ for every $\varphi \in \mathcal{Z}(\mathbb{R}^n)$. Let us show that $\mathcal{F}(f\varphi) \in \mathcal{D}(\mathbb{R}^n)$. This results from the formula

$$T_{\mathcal{F}(f\varphi)} = \mathcal{F}T_f * \mathcal{F}\varphi$$

and the fact that $\mathcal{F}T_f * \mathcal{F}\varphi$ has compact support. As $\mathcal{F}T_f \in \mathcal{S}'(\mathbb{R}^n)$ and $\mathcal{F}\varphi \in \mathcal{D}(\mathbb{R}^n)$, one has

$$\langle \mathcal{F}T_f * \mathcal{F}\varphi, \psi \rangle = \left\langle \mathcal{F}T_f, \overset{\vee}{\mathcal{F}\varphi} * \psi \right\rangle$$
$$= \left\langle T_f, \mathcal{F}(\overset{\vee}{\mathcal{F}\varphi} * \psi) \right\rangle$$
$$= \left\langle T_f, \mathcal{F}\overset{\vee}{\mathcal{F}\varphi}\mathcal{F}\psi) \right\rangle$$
$$= \langle T_f, \varphi\mathcal{F}\psi \rangle$$
$$= \langle T_{\mathcal{F}(f\varphi)}, \psi \rangle, \quad \text{by the Transfer Formula.}$$

Continuity results from the definition of the topology of $\mathcal{Z}(\mathbb{R}^n)$, the formula $T_{\mathcal{F}(f\varphi)} = \mathcal{F}T_f * \mathcal{F}\varphi$ and Proposition 4.9 of Chapter 4.

□

The previous proposition allows a definition of the product of an ultra-distribution T by a function $f \in \mathcal{O}(\mathbb{R}^n)$, as follows

$$\langle fT, \varphi \rangle = \langle T, f\varphi \rangle \; ; \; \varphi \in \mathcal{Z}(\mathbb{R}^n).$$

Indeed, we have $fT \in \mathcal{Z}'(\mathbb{R}^n)$. Moreover the map $T \longmapsto fT$ is continuous from $\mathcal{Z}'(\mathbb{R}^n)$ into $\mathcal{Z}'(\mathbb{R}^n)$, since $p_\varphi(fT) = p_{f\varphi}((T)$ for every $\varphi \in \mathcal{Z}(\mathbb{R}^n)$.

The Exchange Theorem in $\mathcal{S}'(\mathbb{R}^n)$ extends to $\mathcal{Z}'(\mathbb{R}^n)$. In fact, for every $f \in \mathcal{O}(\mathbb{R}^n)$ and $T \in \mathcal{Z}'(\mathbb{R}^n)$, one has

$$\langle \mathcal{F}(fT), \varphi \rangle = \langle fT, \mathcal{F}\varphi \rangle ; \quad \varphi \in \mathcal{D}(\mathbb{R}^n)$$
$$= \langle T, f\mathcal{F}\varphi \rangle$$
$$= \langle \mathcal{F}T, \overline{\mathcal{F}}(f\mathcal{F}\varphi) \rangle$$
$$= \left\langle \mathcal{F}T, b(T_{\overline{\mathcal{F}}(f\mathcal{F}\varphi)}) \right\rangle$$
$$= \langle \mathcal{F}T, b(\overline{\mathcal{F}}T_{f\mathcal{F}\varphi}) \rangle ; \quad \text{by the Transfer Formula in } \mathcal{S}(\mathbb{R}^n)$$
$$= \langle \mathcal{F}T, b(\overline{\mathcal{F}}T_f * \varphi) \rangle ; \quad \text{by Proposition 6.25}$$
$$= \langle \mathcal{F}T, g \rangle,$$

where

$$g(t) = \left\langle (\overline{\mathcal{F}}T_f)_x, \varphi(t-x) \right\rangle$$

$$= \left\langle (\overset{\vee}{\mathcal{F}T_f})_x, \varphi(t-x) \right\rangle$$

$$= \left\langle (\mathcal{F}T_f)_x, \varphi(t+x) \right\rangle.$$

Hence $\langle \mathcal{F}(fT), \varphi \rangle = \langle (\mathcal{F}T)_t, \langle (\mathcal{F}T_f)_x, \varphi(t+x) \rangle \rangle = \langle \mathcal{F}T_f * \mathcal{F}T, \varphi \rangle.$
So we have:

Proposition 6.29. *For every* $f \in \mathcal{O}(\mathbb{R}^n)$ *and any* $T \in \mathcal{Z}'(\mathbb{R}^n)$, *one has*

$$\mathcal{F}(fT) = \mathcal{F}T_f * \mathcal{F}T.$$

Here is an equivalent formulation of the Exchange Theorem.

Proposition 6.30. *For any* $T_1 \in \mathcal{E}'(\mathbb{R}^n)$ *and any* $T_2 \in \mathcal{D}'(\mathbb{R}^n)$, *one has*

$$\overline{\mathcal{F}}(T_1 * T_2) = b((\overline{\mathcal{F}}T_1)\overline{\mathcal{F}}T_2.$$

Proof. Notice first that T_1 is a convolutor, by Proposition 6.19. Hence by Corollary 6.1, one has $T_1 = \mathcal{F}T_f$ with $f \in \mathcal{O}(\mathbb{R}^n)$. On the other hand $\overline{\mathcal{F}}T_2 \in \mathcal{Z}'(\mathbb{R}^n)$. Then by the previous proposition,

$$\mathcal{F}(f\overline{\mathcal{F}}T_2) = \mathcal{F}(T_f) * \mathcal{F}\left(\overline{\mathcal{F}}T_2\right) = T_1 * T_2. \qquad \square$$

Remark 6.19. The Exchange Theorem is deduced from Proposition 6.30. Indeed, for $f \in \mathcal{O}(\mathbb{R}^n)$ and $T \in \mathcal{Z}'(\mathbb{R}^n)$, take $T_1 = \mathcal{F}T_f$ and $T_2 = \mathcal{F}T$.

Remark 6.20. In an analogous manner, one obtains the following identities.

1) $\overline{\mathcal{F}}(fT) = \overline{\mathcal{F}}(T_f) * \overline{\mathcal{F}}T;$ $f \in \mathcal{O}(\mathbb{R}^n)$; $T \in \mathcal{Z}'(\mathbb{R}^n)$.
2) $\mathcal{F}(T_1 * T_2) = b(\mathcal{F}T_1)\mathcal{F}T_2;$ $T_1 \in \mathcal{E}'(\mathbb{R}^n)$; $T_2 \in \mathcal{D}'(\mathbb{R}^n)$.

Remark 6.21. As a consequence of the Exchange Theorem, one gets that $\mathcal{O}(\mathbb{R}^n)$ endowed with the usual addition and product is a commutative and unital algebra on which $\mathcal{Z}(\mathbb{R}^n)$, $\mathcal{Z}'(\mathbb{R}^n)$ and $\mathcal{O}'(\mathbb{R}^n)$ are unital modules.

We are now interested in the convolution product in $\mathcal{Z}(\mathbb{R}^n)$ and its consequences.

It is clear that $\varphi * \psi \in \mathcal{Z}(\mathbb{R}^n)$ for $\varphi, \psi \in \mathcal{Z}(\mathbb{R}^n)$, in view of $\mathcal{F}(\varphi * \psi) = \mathcal{F}\varphi \mathcal{F}\psi$. Moreover for any $\psi \in \mathcal{Z}(\mathbb{R}^n)$, the map $\varphi \longmapsto \varphi * \psi$ of $\mathcal{Z}(\mathbb{R}^n)$ into $\mathcal{Z}(\mathbb{R}^n)$ is continuous since the map $f \longmapsto fg$ of $\mathcal{D}(\mathbb{R}^n)$ into $\mathcal{D}(\mathbb{R}^n)$ is continuous, for any $g \in \mathcal{D}(\mathbb{R}^n)$.

For $T \in \mathcal{S}'(\mathbb{R}^n)$ and $\psi \in \mathcal{S}(\mathbb{R}^n)$, we have defined $\psi * T$ by

$$\langle \psi * T, \varphi \rangle = \left\langle T, \overset{\vee}{\psi} * \varphi \right\rangle, \quad \text{for every } \varphi \in \mathcal{S}(\mathbb{R}^n).$$

But $\left\langle T, \overset{\vee}{\psi} * \varphi \right\rangle$ has a meaning for $T \in \mathcal{Z}'(\mathbb{R}^n)$ and $\psi, \varphi \in \mathcal{Z}(\mathbb{R}^n)$.

Definition 6.13. Let $T \in \mathcal{Z}'(\mathbb{R}^n)$ and $\psi \in \mathcal{Z}(\mathbb{R}^n)$. We put

$$\langle \psi * T, \varphi \rangle = \left\langle T, \overset{\vee}{\psi} * \varphi \right\rangle, \quad \varphi \in \mathcal{Z}(\mathbb{R}^n).$$

Remark 6.22. The map $T \longmapsto \psi * T$ is the transpose of the map $\varphi \longmapsto \overset{\vee}{\psi} * \varphi$ of $\mathcal{Z}(\mathbb{R}^n)$ into $\mathcal{Z}(\mathbb{R}^n)$. Hence it is a continuous map from $\mathcal{Z}'(\mathbb{R}^n)$ into $\mathcal{Z}'(\mathbb{R}^n)$.

Proposition 6.31 (Exchange Theorem). *For any $T \in \mathcal{Z}'(\mathbb{R}^n)$ and any $\psi \in \mathcal{Z}(\mathbb{R}^n)$, one has*

$$\mathcal{F}(\psi * T) = \mathcal{F}\psi \mathcal{F}T \text{ and } \psi * T \in \mathcal{O}_1(\mathbb{R}^n).$$

Proof.   We have

$$\langle \mathcal{F}(\psi * T), \varphi \rangle = \langle \psi * T, \mathcal{F}\varphi \rangle ; \varphi \in \mathcal{D}(\mathbb{R}^n)$$
$$= \left\langle T, \overset{\vee}{\psi} * \mathcal{F}\varphi \right\rangle$$
$$= \left\langle \mathcal{F}T, \overline{\mathcal{F}}(\overset{\vee}{\psi} * \mathcal{F}\varphi) \right\rangle$$
$$= \left\langle \mathcal{F}T, (\overline{\mathcal{F}}\overset{\vee}{\psi})\varphi \right\rangle ; \quad \text{by the Exchange Theorem in } \mathcal{S}(\mathbb{R}^n)$$
$$= \langle \mathcal{F}T, (\mathcal{F}\psi)\varphi \rangle$$
$$= \langle \mathcal{F}\psi \mathcal{F}T, \varphi \rangle.$$

So

$$\mathcal{F}(\psi * T) = \mathcal{F}\psi \mathcal{F}T \text{ and } \mathcal{F}(\psi * T) \in \mathcal{E}'(\mathbb{R}^n).$$

The fact that $\psi * T \in \mathcal{O}_1(\mathbb{R}^n)$ results then from Corollary 6.1 and Proposition 6.19. $\square$

Corollary 6.2. *For any $T \in {}'\mathcal{O}(\mathbb{R}^n)$, the mapping $\psi \longmapsto \psi * T$, of $\mathcal{Z}(\mathbb{R}^n)$ into $\mathcal{Z}_1(\mathbb{R}^n)$ is continuous when $\mathcal{Z}_1(\mathbb{R}^n)$ is endowed with the topology coming from $\mathcal{Z}(\mathbb{R}^n)$ by the isomorphism b.*

Proof. One has $\mathcal{F}T = T_g$ with $g \in \mathcal{E}(\mathbb{R}^n)$. Yet, by the previous proposition, we do have $\mathcal{F}(\psi * T) = T_{g\mathcal{F}\psi}$. Then $\psi * T = T_{\overline{\mathcal{F}}(g\mathcal{F}\psi)}$ with $\overline{\mathcal{F}}(g\mathcal{F}\psi) \in \mathcal{Z}(\mathbb{R}^n)$. Continuity results from the one of $\psi \longmapsto \overline{\mathcal{F}}(g\mathcal{F}\psi)$. For $T \in \mathcal{O}'_c(\mathbb{R}^n)$ and $S \subset \mathcal{S}'(\mathbb{R}^n)$, we have defined $T * S$ by

$$\langle T * S, f \rangle = \left\langle S, b(\overset{\vee}{T} * f) \right\rangle; \quad f \in \mathcal{S}(\mathbb{R}^n).$$

On the other hand $\mathcal{O}'_c(\mathbb{R}^n) \subset {}'\mathcal{O}(\mathbb{R}^n)$, by Corollary 6.1. Now $\left\langle S, b(\overset{\vee}{T} * f) \right\rangle$ is meaningful whenever $T \in {}'\mathcal{O}(\mathbb{R}^n)$ and $S \in \mathcal{Z}'(\mathbb{R}^n)$. $\square$

Definition 6.14. For $T \in {}'\mathcal{O}(\mathbb{R}^n)$ and $S \in \mathcal{Z}'(\mathbb{R}^n)$, we put

$$\langle T * S, \varphi \rangle = \left\langle S, b(\overset{\vee}{T} * \varphi) \right\rangle; \quad \varphi \in \mathcal{Z}(\mathbb{R}^n).$$

Remark 6.23. The map $S \longmapsto T * S$, of $\mathcal{Z}'(\mathbb{R}^n)$ into $\mathcal{Z}'(\mathbb{R}^n)$ is continuous. Indeed for every $\varphi \in \mathcal{Z}(\mathbb{R}^n)$,

$$p_\varphi(T * S) = p_{b(\overset{\vee}{T} * \varphi)}(S), \quad S \in \mathcal{Z}'(\mathbb{R}^n).$$

Notice that it is the transpose of the map $\varphi \longmapsto b(\overset{\vee}{T} * \varphi)$.

Proposition 6.32 (Exchange Theorem). *For any $T \in {}'\mathcal{O}(\mathbb{R}^n)$ and $S \in \mathcal{Z}'(\mathbb{R}^n)$, one has*

$$\mathcal{F}(T * S) = b(\mathcal{F}T)(\mathcal{F}S).$$

Proof. Let $g \in \mathcal{E}(\mathbb{R}^n)$ such that $\mathcal{F}T = T_g$. For every $\varphi \in \mathcal{D}(\mathbb{R}^n)$, one has

$$\begin{aligned}
\langle \mathcal{F}(T * S), \varphi \rangle &= \langle T * S, \mathcal{F}\varphi \rangle \\
&= \left\langle S, b(\overset{\vee}{T} * \mathcal{F}\varphi) \right\rangle \\
&= \langle S, \mathcal{F}(\varphi g) \rangle; \text{ for } \overset{\vee}{T} * \mathcal{F}\varphi = T_{\mathcal{F}(\varphi g)}. \\
&= \langle \mathcal{F}S, g\varphi \rangle \\
&= \langle g\mathcal{F}S, \varphi \rangle \\
&= \langle b(\mathcal{F}T)\mathcal{F}S, \varphi \rangle.
\end{aligned}$$
$\square$

Here is an equivalent formulation of the preceding Exchange Theorem.

Proposition 6.33. *For any* $f \in \mathcal{E}(\mathbb{R}^n)$ *and* $T \in \mathcal{D}'(\mathbb{R}^n)$,

$$\mathcal{F}(fT) = \overline{\mathcal{F}}T_f * \overline{\mathcal{F}}T.$$

Proof.   Since $\overline{\mathcal{F}}T_f \in {}'\mathcal{O}(\mathbb{R}^n)$ and $\overline{\mathcal{F}}T \in \mathcal{Z}'(\mathbb{R}^n)$, one has, by the previous proposition

$$\mathcal{F}(\overline{\mathcal{F}}T_f * \overline{\mathcal{F}}T) = b\left[\mathcal{F}(\overline{\mathcal{F}}T_f)\right]\mathcal{F}\overline{\mathcal{F}}T = fT.$$
$\square$

Remark 6.24. Proposition 6.32 can be deduced from Proposition 6.33. Indeed, for $T_1 \in {}'\mathcal{O}(\mathbb{R}^n)$ and $T_2 \in \mathcal{Z}'(\mathbb{R}^n)$, take $f = b(\mathcal{F}T_1)$ and $T = \mathcal{F}T_2$.

Remark 6.25. In an analogous manner, one obtains the following identities:

(1) $\overline{\mathcal{F}}(T * S) = b(\overline{\mathcal{F}}T)(\overline{\mathcal{F}}S)$, $T \in {}'\mathcal{O}(\mathbb{R}^n)$, $S \in \mathcal{Z}'(\mathbb{R}^n)$.
(2) $\mathcal{F}(fT) = \mathcal{F}T_f * \mathcal{F}T$, $f \in \mathcal{E}(\mathbb{R}^n)$, $T \in \mathcal{D}'(\mathbb{R}^n)$.

Remark 6.26. As a consequence of the Exchange Theorem, one gets that ${}'\mathcal{O}(\mathbb{R}^n)$ is a unital and commutative algebra over which $\mathcal{Z}(\mathbb{R}^n)$, $\mathcal{Z}'(\mathbb{R}^n)$ and $\mathcal{O}(\mathbb{R}^n)$ are unital modules.

The Schwartz space $\mathcal{S}(\mathbb{R}^n)$ and its dual $\mathcal{S}'(\mathbb{R}^n)$ (tempered distributions) are particularly interesting among those appearing in distribution theory. Of course, it is not possible to define the Fourier transform of any distribution $T \in \mathcal{D}'(\mathbb{R}^n)$ by

$$\langle \mathcal{F}T, \varphi \rangle = \langle T, \mathcal{F}\varphi \rangle, \text{ with } \varphi \in \mathcal{D}(\mathbb{R}^n).$$

Indeed, $\mathcal{F}\varphi$ is never in $\mathcal{D}(\mathbb{R}^n)$, for $\varphi \in \mathcal{D}(\mathbb{R}^n) \setminus \{0\}$ (cf. Section 6.1). But it can be done for the subspace $\mathcal{S}'(\mathbb{R}^n)$ of $\mathcal{D}'(\mathbb{R}^n)$. The space $\mathcal{S}(\mathbb{R}^n)$ has been constructed exactly for that purpose (see beginning of Section 6.1). It is actually, an algebra with so many other good properties. The spaces $\mathcal{S}(\mathbb{R}^n)$ and $\mathcal{S}'(\mathbb{R}^n)$ are so basic that no course can omit them. Here all the operations and notions are suggested/justified by first considering functions or regular distributions.

The notion of multiplier raises from the lack of $\mathcal{S}(\mathbb{R}^n)$ to be stable for the multiplication by function of class $\mathcal{C}^\infty$ (see Section 2). The same for the one of a convoluter when the operation is the convolution product. The two notions are intimately related; precisely each convoluter is a Fourier transform of a multiplier.

Now, the formula $\langle \mathcal{F}T, \varphi \rangle = \langle T, \mathcal{F}\varphi \rangle$ has still meaning, whenever $\varphi \in \mathcal{S}(\mathbb{R}^n)$ and $\mathcal{F}\varphi \in \mathcal{D}(\mathbb{R}^n)$. This leads to the introduction of the space

$$\mathcal{Z}(\mathbb{R}^n) = \{ \varphi \in \mathcal{S}(\mathbb{R}^n) : \mathcal{F}\varphi \in \mathcal{D}(\mathbb{R}^n) \}.$$

It appears that $\mathcal{F} : \mathcal{Z}(\mathbb{R}^n) \longrightarrow \mathcal{D}(\mathbb{R}^n)$ is actually an algebraic isomorphism. The method of topology transfer from $\mathcal{D}(\mathbb{R}^n)$ to $\mathcal{Z}(\mathbb{R}^n)$, yields the topological dual $\mathcal{Z}'(\mathbb{R}^n)$ which allows the definition of the Fourier transform of any distribution $T \in \mathcal{D}'(\mathbb{R}^n)$; the same is called the space of ultradistributions. We have included them here for a single reason: to extend the Fourier transformation to the whole space $\mathcal{D}'(\mathbb{R}^n)$. This is developed in [15] in relation with questions in harmonic analysis.

Chapter 7

Structure of Distributions

In this chapter, we present two types of structure theorems. The first, called local, describes any distribution on every $\mathcal{D}_K(\Omega)$ with K a compact subset of Ω. The second, said global, considers distributions with compact support. In the case of distributions with singletons as supports, one has a structure result in terms of Dirac distributions. Finally, we also give a structure theorem for tempered distributions.

7.1 Local structure

In this section, we consider the case of an arbitrary distribution.

Proposition 7.1 (Local structure of distributions). *Let Ω be an open subset of $\mathbb{R}^n$, $T \in \mathcal{D}'(\Omega)$ and K a compact set in Ω. Then there is a continuous function f on Ω and a positive integer m such that*

$$(1) \qquad T = \frac{\partial^{mn}}{\partial x_1^m \cdots \partial x_n^m} T_f \ \text{on} \ \mathcal{D}_K(\Omega).$$

Moreover the support of f can be chosen in an arbitrary neighborhood of K.

Proof. We will first show the existence of a function $f \in L^\infty(K)$ satisfying (1), i.e., for every $\varphi \in \mathcal{D}_K(\Omega)$

$$(1') \qquad \langle T, \varphi \rangle = (-1)^{mn} \int_{\mathbb{R}^n} f(x) \frac{\partial^{mn}}{\partial x_1^m \cdots \partial x_n^m} \varphi(x) \ dx.$$

If we put $c = \|f\|_\infty$ the norm of f in $L^\infty(K)$, then the relation $(1')$ gives

$$(2) \qquad |\langle T, \varphi \rangle| \leq c \int_{\mathbb{R}^n} \left| \frac{\partial^{mn}}{\partial x_1^m \cdots \partial x_n^m} \varphi(x) \right| dx, \ \varphi \in \mathcal{D}_K(\Omega).$$

Actually, the latter relation gives (1'). Indeed suppose we have a constant $c > 0$ such that (2) is satisfied. With every $\varphi \in \mathcal{D}_K(\Omega)$, associate the function ψ defined on Ω by

$$\psi(x) = (-1)^{mn} \frac{\partial^{mn}}{\partial x_1^m \cdots \partial x_n^m} \varphi(x).$$

It is clear that $\psi \in L^1(K)$. Moreover the linear map $\varphi \longmapsto \psi$ is one-to-one. Indeed if ψ is zero, then φ is also zero by successive integrations. Define, on the vector subspace $E = \{\psi : \varphi \in \mathcal{D}_K(\Omega)\}$ of $L^1(K)$, a linear form by $\psi \longmapsto \langle T, \varphi \rangle$. It is continuous by the relation (2), so that by the Hahn-Banach theorem, it extends to a continuous linear form on $L^1(K)$. As $L^\infty(K)$ is the topological dual of $L^1(K)$, there is a function $f \in L^\infty(K)$ such that $\|f\|_\infty \leq c$ and

$$\langle T, \varphi \rangle = \int_{\mathbb{R}^n} \psi(x) f(x) dx$$

which is exactly (1'). So it is sufficient to show (2). As $T \in \mathcal{D}'(\Omega)$, there is a constant $c' > 0$ and an integer k such that

$$(3) \qquad |\langle T, \varphi \rangle| \leq c' \sup_{\substack{x \in K \\ |\alpha| \leq k}} |D^\alpha \varphi(x)| \, ; \, \varphi \in \mathcal{D}_K(\Omega).$$

Put $a_j = \sup_{x \in K} |x_j|$, $j = 1, 2, \ldots, n$. For every $\psi \in \mathcal{D}_K(\Omega)$, one has by the mean value theorem

$$\sup_{x \in K} |\psi(x)| \leq a_j \sup_{x \in K} \left| \frac{\partial}{\partial x_j} \psi(x) \right|.$$

Applying repeatedly this inequality, one obtains from (3)

$$(4) \qquad |\langle T, \varphi \rangle| \leq c \sup_{x \in K} \left| \frac{\partial^k}{\partial x_1^k} \cdots \frac{\partial^k}{\partial x_n^k} \varphi(x) \right|, \quad \varphi \in \mathcal{D}_K(\Omega),$$

where $c > 0$ is a suitably chosen constant. But if $\psi \in \mathcal{D}_K(\Omega)$, one has for every $i = 1, 2, \ldots, n$,

$$\psi(y_1, \ldots, y_n) = \int_{-\infty}^{y_i} \frac{\partial \psi}{\partial x_i} (y_1, y_2, \ldots, y_{i-1}, x_i, y_{i+1}, \ldots, y_n) dx_i,$$

whence

$$\psi(y) = \int_{-\infty}^{y_1} \cdots \int_{-\infty}^{y_n} \frac{\partial}{\partial x_1} \cdots \frac{\partial}{\partial x_n} \psi(x) dx.$$

Hence

$$(5) \qquad \sup_{x \in K} |\psi(y)| \leq \int_{\mathbb{R}^n} \left| \frac{\partial}{\partial x_1} \cdots \frac{\partial}{\partial x_n} \psi(x) \right| dx.$$

Taking

$$\psi = \frac{\partial^k}{\partial x_1^k} \cdots \frac{\partial^k}{\partial x_n^k} \varphi$$

and combining (4) and (5), one obtains

$$|\langle T, \varphi \rangle| \leq c \int_{\mathbb{R}^n} \left| \frac{\partial^{(k+1)n}}{\partial x_1^{k+1} \cdots \partial x_n^{k+1}} \varphi(x) \right| dx.$$

Whence (2) with $m = k + 1$. Now, to obtain the relation (1), extend f to $\mathbb{R}$ by zero outside of K. One has $f \in L^\infty(\mathbb{R}^n)$. Put

$$F(x_1, ..., x_n) = \int_{-\infty}^{x_1} \cdots \int_{-\infty}^{x_n} f(y_1, ..., y_n) dy_1 \cdots dy_n.$$

Then $F \in L^1(\mathbb{R}^n)$. Hence, on one hand F is continuous by Lebesgue's theorem. On the other hand, we have

$$\frac{\partial^n}{\partial x_1 \cdots \partial x_n} F = f, \quad a.e..$$

Therefore

$$\frac{\partial^n}{\partial x_1 \cdots \partial x_n} T_F = T_f, \text{ since } F \text{ is continuous.}$$

Hence

$$T = \frac{\partial^{(m+1)n}}{\partial x_1^{m+1} \cdots \partial x_n^{m+1}} T_F.$$

Finally if U is an arbitrary neighborhood of K then, by a lemma of Urysohn type (cf. [15]), there is a function χ of class $\mathcal{C}^\infty$ on $\mathbb{R}$ with a support contained in U, such that $\chi = 1$ on K. Consider then the function χF. $\square$

Remark 7.1. Let $T \in \mathcal{D}'(\mathbb{R}^n)$ and Ω be a relatively compact subset of $\mathbb{R}^n$. The previous result shows that the restriction of T to $\mathcal{D}(\Omega)$ is of finite order.

7.2 Global structure

For the distributions with compact support, one has a global structure result.

Proposition 7.2. (Global structure of a distribution with a compact support). *Let Ω be an open subset of $\mathbb{R}^n$ and T a distribution with* compact support K in Ω. Then, there is a finite number of continuous functions f_q, $q \in \mathbb{N}^n$, on Ω such that

$$T = \sum_q D^q T_{f_q}, \quad \text{on } \mathcal{D}(\Omega).$$

Moreover, the supports of the f_q's can be chosen in an arbitrary neighborhood of K.

Proof. Let U be an arbitrary neighborhood of K and V a relatively compact neighborhood of K such that $K \subset V \subset \overline{V} \subset U \subset \Omega$. According to the previous proposition, there is a continuous function f with support contained in U and a multi-index $p \in \mathbb{N}^n$ such that

$$T = D^p T_f, \quad \text{on } \mathcal{D}_{\overline{V}}(\Omega).$$

Now, let $\varphi \in \mathcal{E}(\Omega)$ and $\chi_0 \in \mathcal{D}(\Omega)$, $\chi_0 = 1$ on a neighborhood of K with $supp \chi_0 \subset V$. One has $supp\ \chi_0 \varphi \subset V$ and

$$\langle T, \varphi \rangle = \langle T, \chi_0 \varphi \rangle = (-1)^{|p|} \int_{\mathbb{R}^n} f(x) D^p (\chi_0 \varphi)(x) dx.$$

But by the Leibniz formula

$$D^p(\chi_0 \varphi) = \sum_{q \le p} c(p,q) D^{p-q} \chi_0 D^q \varphi,$$

where $c(p,q) = \frac{p!}{q!(p-q)!}$. Hence

$$\langle T, \varphi \rangle = (-1)^{|p|} \sum_{q \le p} C(p,q) \int_{\mathbb{R}^n} f(x) D^{p-q} \chi_0(x) D^q \varphi(x) dx$$

$$= (-1)^{|p|} \sum_{q \le p} C(p,q)(-1)^{|q|} \left\langle D^q T_{f D^{p-q} \chi_0}, \varphi \right\rangle$$

$$= \left\langle \sum_{q \le p} D^q T_{f_q}, \varphi \right\rangle,$$

where $f_q = (-1)^{|p+q|} c(p,q)\ f\ D^{p-q} \chi_0$; f_q is continuous and $supp\ f_q \subset U$ for every $q \in \mathbb{N}^n$.

$\square$

The supports of the distributions $D^p \delta_a$, $p \in \mathbb{N}^n$, are all reduced to $\{a\}$. It is also so for any finite linear combination of them. The converse is valid as we will now show.

Proposition 7.3. (Structure of distributions with singletons as supports). *Let $T \in \mathcal{D}'(\mathbb{R}^n)$ such that $supp\, T = \{a\}$. Then, there is $m \in \mathbb{N}^n$ and complex constants c_p, $p \in \mathbb{N}^n$, $|p| \le m$ so that*

$$T = \sum_{|p| \le m} c_p D^p \delta_a.$$

Proof. Replacing T by $\tau_{-a}T$ if necessary, we can suppose that $supp\, T = \{0\}$. Let $\varphi \in \mathcal{E}(\mathbb{R}^n)$. By Taylor expansion one has for every $m \in \mathbb{N}$

$$\varphi(x) = \sum_{|p| \leq m} \frac{D^p\varphi(0)}{p!}\, x^p + \varphi_m(x),$$

where $\varphi_m \in \mathcal{E}(\mathbb{R}^n)$ is such that all its derivatives of order $\leq m$ are zero at the origin. One then has

$$\langle T, \varphi \rangle = \sum_{|p| \leq m} \frac{\langle T, x^p \rangle}{p!}\, D^p\varphi(0) + \langle T, \varphi_m \rangle$$

$$= \left\langle \sum_{|p| \leq m} C_p\, D^p\delta, \varphi \right\rangle + \langle T, \varphi_m \rangle,$$

where

$$c_p = \frac{(-1)^{|p|}\, \langle T, x^p \rangle}{p!}.$$

To finish, it suffices to show the existence of an m such that $\langle T, \varphi_m \rangle = 0$ for every φ. This follows from the following general result. $\square$

Proposition 7.4. *Let T be a distribution on Ω with compact support K. Then, there is $m \in \mathbb{N}$ such that $\langle T, \varphi \rangle = 0$, for every $\varphi \in \mathcal{E}(\Omega)$ which vanishes on K as well as all its derivatives of order $\leq m$.*

Proof. The distribution T, having compact support, is of finite order m. We will show that for any suitable $\varepsilon > 0$, φ is equal on a neighborhood V_ε of $K = supp\, T$ to a function $\varphi_\varepsilon \in \mathcal{E}(\Omega)$ such that (φ_ε) tends to zero, as well as its derivatives of order $\leq m$, uniformly on Ω when $\varepsilon \longrightarrow 0$. Let us look for φ_ε under the form $\psi_\varepsilon\varphi$ where $\psi_\varepsilon \in \mathcal{D}(\Omega)$ is equal to 1 on a neighborhood of K. Consider δ such that $0 < \delta < d(K, \Omega^c)$ and $K_\varepsilon = K + \overline{B}(0, \varepsilon)$ with $0 < \varepsilon < \delta/4$. Take $\psi_\varepsilon = \theta_\varepsilon * \chi_{K_{2\varepsilon}}$ the regularization of the characteristic function of $K_{2\varepsilon}$. It is known that $\psi_\varepsilon \in \mathcal{D}(\Omega)$ with $supp\, \psi_\varepsilon \subset K_{3\varepsilon}$, $0 \leq \psi_\varepsilon \leq 1$ and $\psi_\varepsilon = 1$ on $K_{2\varepsilon}$. Moreover

$$(1) \qquad \left| D^\beta\psi_\varepsilon(x) \right| \leq \frac{1}{\varepsilon^{|\beta|}} \int_{\mathbb{R}^n} \left| D^\beta\theta(x) \right| dx, \quad |\beta| \leq m.$$

On the other hand, for $y \in K$ and $y + h \in K_\delta$ one has by Taylor's formula

$$\varphi(y + h) = (m+1) \sum_{|\beta| = m+1} \frac{h^\beta}{\beta!} \int_0^1 (1 - t)^m D^\beta\varphi(y + t\, h)dt,$$

whence $|\varphi(y+h)| \leq c_1 |h|^{m+1}$, where c_1 is a constant independent of δ, φ and m. An analogous inequality for $D^\gamma \varphi(y+h)$, $|\gamma| \leq m$, gives

$$(2) \qquad \sup_{x \in K_{3\varepsilon}} |D^\gamma \varphi(x)| \leq c_2 \, \varepsilon^{m+1-|\gamma|}, \quad |\gamma| \leq m$$

where c_2 is a constant which is also independent of δ, φ and m. Notice that $supp\,(1 - \psi_\varepsilon)\,\varphi$ is contained in the zero open set of T. Hence

$$(3) \qquad \langle T, \varphi \rangle = \langle T, \psi_\varepsilon \, \varphi \rangle \,.$$

As T is of order $\leq m$, there is a constant c_3 such that

$$|\langle T, \psi_\varepsilon \, \varphi \rangle| \leq c_3 \sup_{\substack{x \in K_{3\varepsilon} \\ |\alpha| \leq m}} |D^\alpha(\psi_\varepsilon \, \varphi)(x)| \,.$$

Finally, by Leibniz formula and the inequalities (1), (2) and (3), one has

$$|\langle T, \varphi \rangle| \leq c_4 \, \varepsilon$$

where c_4 is a constant not depending of ε. $\square$

Remark 7.2. If $T \in \mathcal{D}'(\Omega)$ and $\varphi \in \mathcal{D}(\Omega)$ are such that $\varphi = 0$ on $supp\,T$, one does not necessarily have $\langle T, \varphi \rangle = 0$. Take for example $T = \delta'$ on $\mathbb{R}$ and a function $\varphi \in \mathcal{D}(\mathbb{R})$ such that $\varphi(x) = x$ on $[-1, 1]$.

Proposition 7.5. *Let $T \in \mathcal{D}'(\Omega)$ have compact support and order less than or equal to m. If $(\varphi_j)_j \subset \mathcal{D}(\Omega)$ converges uniformly to 0 on a neighborhood V of $supp\,T$, as well as all its derivatives of order less or equal to m, then $\langle T, \varphi_j \rangle$ tends to zero.*

Proof. Let $\chi \in \mathcal{D}(\Omega)$ equal to 1 on a compact neighborhood L of $supp\,T$ with $L \subset V$. The functions $\psi_j = \chi \varphi_j$ in $\mathcal{D}(\Omega)$ have their supports in $supp\,\psi$, and $\langle T, \psi_j \rangle = \langle T, \varphi_j \rangle$. But (ψ_j) converges to 0 in $\mathcal{D}_L(\Omega)$, hence $\langle T, \psi_j \rangle \xrightarrow[j]{} 0$. $\square$

Remark 7.3. Uniform convergence on $supp\,T$ is not sufficient in general. Indeed, the distribution T defined on $\mathbb{R}$ by

$$\langle T, \varphi \rangle = \lim_{m \to +\infty} \left[\left(\sum_{p=1}^{m} \varphi(\tfrac{1}{p}) \right) - m\,\varphi(0) - (\log\,m)\,\varphi'(0) \right]$$

has compact support $K = \left\{ \dfrac{1}{p} : p \in \mathbb{N}^* \right\} \cup \{0\}$. For $j \in \mathbb{N}^*$, consider $\varphi_j \in \mathcal{D}(\mathbb{R})$ given by

$$\varphi_j(x) = \begin{cases} \frac{1}{\sqrt{j}} & \text{if } x \geq \frac{1}{j}, \\ 0 & \text{if } x \leq \frac{1}{j+1}. \end{cases}$$

It is clear that $(\varphi_j)_j$ satisfies the hypotheses of Proposition 7.5. However $\langle T, \varphi_j \rangle = \sqrt{j}$.

The result of Proposition 7.4 extends to distributions of finite order.

Proposition 7.6. *Let $T \in \mathcal{D}'(\Omega)$ of order less or equal to m. Then $\langle T, \varphi \rangle = 0$, for every $\varphi \in \mathcal{D}(\Omega)$ which vanishes on $supp\,T$ as well as all its derivatives of order less or equal to m.*

Proof. Let $\psi \in \mathcal{D}(\Omega)$ equal to 1 on $supp\,\varphi$. The distribution ψT has compact support and one has

$$\langle \psi T, \varphi \rangle = \langle T, \varphi \rangle.$$

Conclude by Proposition 7.4 in view of $supp\ \psi T \subset supp\ T$.

□

Proposition 7.1 describes the structure of a distribution $T \in \mathcal{D}'(\Omega)$ on a relatively compact subset of Ω. Distributions of finite order show that the result is not valid for an arbitrary open subset. Here is what can be said in the general case.

Proposition 7.7. *Let $T \in \mathcal{D}'(\Omega)$ and $(\Omega_i)_{i \in I}$ be an open covering of $supp\,T$. Then, there is a family $(T_i)_{i \in I}$ of distributions on Ω such that*

(a) $supp\ T_i \subset \Omega_i \cap supp\ T$, for every i.
(b) For every $\varphi \in \mathcal{D}(\Omega)$, $\langle T, \varphi \rangle = \sum\limits_{finite} \langle T_i, \varphi \rangle$. We globally write $T = \sum\limits_i T_i$.

Proof. If we put $\Omega_0 = \Omega \backslash supp\ T$, then $(\Omega_i)_i \cup \{\Omega_0\}$ is a covering of Ω. Let $(g_i)_i$ be the associated locally finite C^∞-partition of unity and put $T_i = g_i T$. It is a distribution with support contained in $\Omega_i \cap supp\,T$. And as on any compact set in Ω only a finite number of g_i's are nonzero, one has for every $\varphi \in \mathcal{D}(\Omega)$

$$\langle T, \varphi \rangle = \sum\limits_{finite} \langle T_i, \varphi \rangle.$$

□

Remark 7.4.

(1) Let $T \in \mathcal{D}'(\Omega)$ and let Ω_i be relatively compact sets such that $\cup \Omega_i \supset supp\,T$. Then, by the previous proposition and Proposition 7.2, there are continuous functions f_i on Ω and positive integers p_i such that

$$T = \sum\limits_i D^{p_i} T_{f_i}.$$

Moreover, the support of each f_i can be chosen in an arbitrary neighborhood of $supp\ T \cap \overline{\Omega_i}$, hence in an arbitrary neighborhood of $supp\ T$.

(2) Let $T \in \mathcal{D}'(\Omega)$. Taking the Ω_i's relatively compact, Propositions 7.4 and 7.5 show that $\langle T, \varphi \rangle = 0$ for every $\varphi \in \mathcal{D}(\Omega)$ null on *supp* T as well as its derivatives of any order.

7.3   Structure of tempered distributions

For tempered distributions we also have a global structure result.

As $\mathcal{S}'(\mathbb{R}^n)$ is stable by derivation, any derivative of a distribution defined by a slowly increasing continuous function is a tempered distribution. We will see that in fact this turns out to be a characterization.

Proposition 7.8. *For any distribution T on $\mathbb{R}^n$, the following assertions are equivalent.*

(i) $T \in \mathcal{S}'(\mathbb{R}^n)$.
(ii) There is a continuous function f slowly increasing on $\mathbb{R}^n$ and a positive integer m such that

$$T = \frac{\partial^{mn} T_f}{\partial x_1^m \cdots \partial x_n^m} \quad on \ \mathbb{R}^n.$$

Proof. $(i) \implies (ii)$ It is sufficient to show the existence of a function $f \in L^2(\mathbb{R}^n)$ and of an integer $k \in \mathbb{N}$ such that

$$T = \frac{\partial^{mn} T_g}{\partial x_1^m \cdots \partial x_n^m}, \text{ where } g(x) = (1 + \|x\|^2)^k f(x)$$

i.e., for every $\varphi \in \mathcal{D}(\mathbb{R}^n)$

$$\langle T, \varphi \rangle = (-1)^{mn} \int_{\mathbb{R}^n} f(x) \left(1 + \|x\|^2\right)^k \frac{\partial^{mn} \varphi(x)}{\partial x_1^m \cdots \partial x_n^m} dx. \tag{7.3.1}$$

Indeed, putting

$$F(x) = \int_0^{x_1} \cdots \int_0^{x_n} \left(1 + s_1^2 + \cdots + s_n^2\right)^k f(s_1, \ldots, s_n) \, ds_1 \ldots ds_n,$$

we have F continuous on $\mathbb{R}^n$. Moreover, by the Schwarz inequality

$$|F(x)|^2 \le \|f\|_2^2 \left(\sum_{|\alpha| \le 2k} \frac{(2k)!}{\alpha!(2k - |\alpha|)!} \frac{|x_1|^{2\alpha_1 + 1}}{2\alpha_1 + 1} \cdots \frac{|x_n|^{2\alpha_n + 1}}{2\alpha_n + 1} \right).$$

This shows that F is slowly increasing, and

$$\frac{\partial^n}{\partial x_1 \cdots \partial x_n} T_F = T_{(1 + \|x\|^2)^k f}.$$

Let us now prove (7.3.1). If we put $c = \|f\|_2$, the norm of f in $L^2(\mathbb{R}^n)$, the relation (7.3.1) gives

$$\langle T, \varphi \rangle \le c \left(\int_{\mathbb{R}^n} \left| \left(1 + \|x\|^2\right)^k \frac{\partial^{mn} \varphi(x)}{\partial x_1^m \cdots \partial x_n^m} \right|^2 dx \right)^{\frac{1}{2}}. \qquad (7.3.2)$$

Conversely suppose that one has (7.3.2). With every $\varphi \in \mathcal{D}(\mathbb{R}^n)$, associate the function

$$\psi(x) = (-1)^{mn} \left(1 + \|x\|^2\right)^k \frac{\partial^{mn} \varphi(x)}{\partial x_1^m \cdots \partial x_n^m}.$$

The map

$$J : (\mathcal{D}(\mathbb{R}^n), Q_{k,m}) \longrightarrow \left(L^2(\mathbb{R}^n), \|\cdot\|_2\right)$$
$$\varphi \longmapsto \psi$$

where

$$Q_{k,m}(\varphi) = \left\| \left(1 + \|x\|^2\right)^k \frac{\partial^{mn} \varphi(x)}{\partial x_1^m \cdots \partial x_n^m} \right\|_2$$

is an isometry. Its transpose

$$t_J : \left(L^2(\mathbb{R}^n), \|\cdot\|_2\right) \longrightarrow (\mathcal{D}(\mathbb{R}^n), Q_{k,m})'$$

is onto. By (7.3.2), $T \in (\mathcal{D}(\mathbb{R}^n), Q_{k,m})'$. Therefore, there is $f \in L^2(\mathbb{R}^n)$ such that $t_J(f) = T$. This is exactly the relation (7.3.1). So (7.3.1) and (7.3.2) are equivalent. One concludes by the following lemma. $\square$

Lemma 7.1. *The family of seminorms* $(Q_{k,m})_{k,m}$ *defines the topology of* $\mathcal{S}(\mathbb{R}^n)$.

Proof. We will show that the family $(Q_{k,m})_{k,m}$ defines the same topology as that defined by $\left(|\cdot|_{k,m}\right)$, where

$$|\varphi|_{k,m} = \max_{|\beta| \le m} \sup_{x \in \mathbb{R}^n} \left| \left(1 + \|x\|^2\right)^k D^m \varphi(x) \right|, \quad \varphi \in \mathcal{S}(\mathbb{R}^n)$$

with

$$D = \frac{\partial^n}{\partial x_1 \cdots \partial x_n}.$$

On one hand

$$Q_{k,m}(\varphi) \le \pi^2 |\varphi|_{k+m,nm}, \quad \text{for every } \varphi \in \mathcal{S}(\mathbb{R}^n),$$

whence the continuity of the seminorms $Q_{k,m}$ on $\mathcal{S}(\mathbb{R}^n)$. On the other hand, we will show that for every $(k_0, m_0) \in \mathbb{N} \times \mathbb{N}$, there is a constant $c > 0$ and $(k, m) \in \mathbb{N} \times \mathbb{N}$ such that

$$|\varphi|_{k_0, m_0} \leq c Q_{k,m}(\varphi), \text{ for every } \varphi \in \mathcal{S}(\mathbb{R}^n).$$

This is derived from the following inequalities: for every $\psi \in \mathcal{S}(\mathbb{R}^n)$,

$$\|\psi\|_{L^1} \leq \pi^n \left\| (1 + \|x\|^2)^n \psi \right\|_{L^2} \tag{7.3.3}$$

$$\left\| (1 + \|x\|^2)^k \psi \right\|_{L^\infty} \leq \left\| (1 + \|x\|^2)^k D\psi \right\|_{L^1} \tag{7.3.4}$$

$$\left\| (1 + \|x\|^2)^k \psi \right\|_{L^\infty} \leq \pi/2 \left\| (1 + \|x\|^2)^{nk+1} \partial_i \psi \right\|_{L^\infty}, \quad i = 1, ..., n \tag{7.3.5}$$

with $m = m_0 + 1$, $c = \pi^{nm}$, $k = n + k_{nm}$ where $k_j = 1 + n k_{j-1}$ and $j = 1, ..., nm$. Indeed, using (7.3.5), taking the derivative $m_0 - \beta_1$ times with respect to $x_1, ...,$ and $m_0 - \beta_n$ times with respect to x_n, one finds

$$\left\| \left(1 + \|x\|^2\right)^k D^\beta \varphi \right\|_\infty \leq (\tfrac{\pi}{2})^{nm_0 - |\beta|} \left\| \left(1 + \|x\|^2\right)^{k_{nm}} D^{(m_0 ... m_0)} \varphi \right\|_\infty.$$

We then use (7.3.4) and next (7.3.3). Now we show the previous three inequalities. The first follows from the Schwarz inequality. For (7.3.4), put $A = \{x = (x_1, ..., x_n) \in \mathbb{R}^n, \ x_1, ..., x_p \geq 0 \text{ and } x_{p+1}, ..., x_n \leq 0\}$. For $x \in A$, one has

$$\psi(x) = (-1)^p \int_{x_1}^{+\infty} \cdots \int_{x_p}^{+\infty} \int_{-\infty}^{x_{p+1}} \cdots \int_{-\infty}^{x_n} D\psi(s_1, ..., s_n) \, ds_1, ..., ds_n.$$

Since $0 \leq x_i \leq s_i$, $i = 1, ..., p$ and $s_j \leq x_j \leq 0$, $j = p+1, ..., n$, we have

$$(1 + \|x\|^2)^k \leq (1 + \|s\|^2)^k, \quad s \in A.$$

Hence, for every $x \in A$

$$\left(1 + \|x\|^2\right)^k |\psi(x)| \leq \int_A \left(1 + \|s\|^2\right)^k |D\psi(s)| \, ds$$

$$\leq \int_{\mathbb{R}^n} \left(1 + \|s\|^2\right)^k |D\psi(s)| \, ds,$$

from where the result follows since $\mathbb{R}^n$ is a union (even finite) of such subsets A.

We will show (7.3.5) for $i = 1$. Writing

$$\psi(x) = -\int_{-\infty}^{x_1} \partial_1 \psi(s_1, x_2, ..., x_n) \, ds_1, \text{ for } x_1 \leq 0$$

and

$$\psi(x) = - \int_{x_1}^{+\infty} \partial_1 \psi(s_1, x_2, ..., x_n) \, ds_1, \text{ for } x_1 \geq 0,$$

one obtains for every $x_1 \in \mathbb{R}$

$$\left(1 + \|x_1\|^2\right)^k |\psi(x)| \leq \frac{\pi}{2} \sup_{t_1 \in \mathbb{R}} \left(1 + \|t_1\|^2\right)^{k+1} |\partial_1 \psi(t_1, x_2, ..., x_n)|.$$

Then

$$\left(1 + \|x\|^2\right)^k |\psi(x)| \leq \frac{\pi}{2} \prod_{i=2}^{n} \left(1 + \|x_i\|^2\right)^k \sup_{x_1 \in \mathbb{R}} \left(1 + \|x_1\|^2\right)^{k+1} |\partial_1 \psi(x_1, ...x_n)|$$

$$\leq \frac{\pi}{2} \sup_{x \in \mathbb{R}} \left(1 + \|x\|^2\right)^{nk+1} |\partial_1 \psi(x)|.$$

$\square$

For $T \in \mathcal{D}'(\Omega)$ one wants to describe the structure of T on $\mathcal{D}_K(\Omega)$, for any compact subset K of Ω; this is what is meant by a local structure. When T can be described on the whole space $\mathcal{D}(\Omega)$, then one knows its global structure. It is worthwhile to point out that these are given in terms of regular distributions or Dirac measures.

To the extent we know the above has never been the matter of a separate chapter. In [6] the results are presented together with the notion of a support. The global structure is given only for distributions with compact support. In the first volume of [15] the global structure in the case of a compact support is left as an exercise. The local structure, is first given as an exercise in the same volume, while it is reconsidered along with the global one of tempered distributions, in the second volume. The proof is based on considerations about duals of general topological vector spaces.

We have chosen to put a separate chapter in order to collect together the different results and to make it clear that the structure has its own interest. It is better to tackle that aspect once the other notions have been independently assimilated.

The proofs go more or less along the lines of those in [6] and [15]. However, they are worked on to make them accessible to the reader. No detail is omitted. We also include results which are often given as exercises. Concerning tempered distributions, the proof of [15] is given here in its proper context without referring to non necessary general results. For more readings, see [6] and [15].

Chapter 8

Fourier Ranges of Some Subspaces of $\mathcal{S}'(\mathbb{R}^n)$

We first present the Plancherel-Parseval theorem in $\mathcal{S}(\mathbb{R}^n)$, then that of Riesz-Parseval in $L^2(\mathbb{R}^n)$. The latter allows a characterization of the Fourier range of $L^1(\mathbb{R}^n)$. Next we describe the Fourier range of regular distributions defined by multipliers. The Paley-Wiener theorem characterizes the one of $\mathcal{D}(\mathbb{R}^n)$. Finally the Fourier range of the space $\mathcal{E}'(\mathbb{R}^n)$ is given by Paley-Wiener-Schwartz theorem.

8.1 Fourier range of $L^2(\mathbb{R}^n)$

The space $\mathcal{S}(\mathbb{R}^n)$ is contained in $L^2(\mathbb{R}^n)$. So we can endow it with a pre-Hilbertian structure.

Proposition 8.1 (Plancherel-Parseval theorem). *The Fourier transformation is an isometry of $\mathcal{S}(\mathbb{R}^n)$ onto $\mathcal{S}(\mathbb{R}^n)$, for the norm of $L^2(\mathbb{R}^n)$, i.e.,*

$$\left\|\hat{f}\right\|_2 = \|f\|_2, \text{ for every } f \in \mathcal{S}(\mathbb{R}^n).$$

Proof. In fact there is even a preservation of the scalar product, i.e.,

$$\left\langle \hat{f}, \hat{g} \right\rangle = \langle f, g \rangle, \text{ for any } f, g \text{ in } \mathcal{S}(\mathbb{R}^n).$$

Indeed

$$\langle \hat{f}, \hat{g} \rangle = \int_{\mathbb{R}^n} (\mathcal{F}f)(y)\overline{\mathcal{F}g}(y)dy$$
$$= \int_{\mathbb{R}^n} (\mathcal{F}f)(y)(\overline{\mathcal{F}g})(y)dy$$
$$= \int_{\mathbb{R}^n} f(y)\mathcal{F}(\overline{\mathcal{F}g})(y)dy; \text{ by the transfer formula.}$$
$$= \langle f, g \rangle.$$
□

The space $S(\mathbb{R}^n)$ is dense in $L^2(\mathbb{R}^n)$, since $\mathcal{D}(\mathbb{R}^n)$ is dense in $L^2(\mathbb{R}^n)$ (cf. Chapter 1). Using the principle of extending by uniform continuity, one has the following result.

Proposition 8.2 (Riesz-Plancherel theorem). *The Fourier transformation extends to an isometry of $L^2(\mathbb{R}^n)$ onto itself.*

Remark 8.1. The previous results are also valid for the Fourier cotransformation $\overline{\mathcal{F}}$. If we still denote by $\overline{\mathcal{F}}$ and $\mathcal{F}$ their respective extensions to $L^2(\mathbb{R}^n)$, we do have $\mathcal{F}\overline{\mathcal{F}} = \overline{\mathcal{F}}\mathcal{F} = Id_{L^2(\mathbb{R}^n)}$ too.

Remark 8.2. If $f \in L^1(\mathbb{R}^n) \cap L^2(\mathbb{R}^n)$, then $\mathcal{F}_1 f = \mathcal{F}_2 f$ where $\mathcal{F}_1$ is the Fourier transformation in $L^1(\mathbb{R}^n)$ and $\mathcal{F}_2$ its extension by Proposition 8.2. Indeed, using the equality of $\mathcal{F}_1$ and $\mathcal{F}_2$ on $S(\mathbb{R}^n)$, the transfer and Plancherel-Parseval formulas, one has

$$\langle \mathcal{F}_1 f, \varphi \rangle = \langle \mathcal{F}_2 f, \varphi \rangle, \text{ for every } \varphi \in S(\mathbb{R}^n).$$

This proves the assertion by Proposition 2.4 of Chapter 2, for $\mathcal{F}_1 f \in L^1_{loc}(\mathbb{R}^n)$ and $\mathcal{F}_2 f \in L^1_{loc}(\mathbb{R}^n)$, since $f \in L^1(\mathbb{R}^n) \cap L^2(\mathbb{R}^n)$.

The extension in Proposition 8.2 is not expressible as an integral like the Fourier transformation in $L^1(\mathbb{R}^n)$. However one has the following result of Riesz-Plancherel:

Proposition 8.3. *Let $f \in L^2(\mathbb{R}^n)$. Put*

$$f_k(y) = \int_{B(0,k)} f(x)e^{-2i\pi xy}dx, \text{ for every } y \in \mathbb{R}^n.$$

Then $\mathcal{F}f$ is the limit in $L^2(\mathbb{R}^n)$ of the sequence $(f_k)_k$. We write for every $y \in \mathbb{R}^n$,

$$\hat{f}(y) = \lim_{k \to +\infty} \int_{B(0,k)} f(x)e^{-2i\pi xy}dx, \text{ in } L^2(\mathbb{R}^n).$$

Proof. Write

$$f_k(y) = \int_{\mathbb{R}^n} g_k(x)e^{-2i\pi xy}dx, \text{ where } g_k(x) = \chi_{B(0,k)}(x)f(x).$$

As $f \in L^2(\mathbb{R}^n)$, one has $g_k \in L^1(\mathbb{R}^n)$ and then $f_k = \hat{g}_k$. The sequence $(g_k)_k$ converges pointwise to f and $|g_k|^2 \leq |f|^2$. By Lebesgue's theorem, the sequence $(g_k)_k$ converges to f in $L^2(\mathbb{R}^n)$. The result then follows from Proposition 8.2 and Remark 8.2. □

In $L^2(\mathbb{R}^n)$, the analogue of (vii) in Proposition 5.1 of Chapter 5 is the following:

Proposition 8.4. *If $f, g \in L^2(\mathbb{R}^n)$, then*

$$\mathcal{F}(fg) = \mathcal{F}f * \mathcal{F}g.$$

Proof. Indeed,

$$
\begin{aligned}
(\mathcal{F}f * \mathcal{F}g)(x) &= \int_{\mathbb{R}^n} \mathcal{F}f(x-y)\mathcal{F}g(y)dy \\
&= \int_{\mathbb{R}^n} \tau_{-x}(\mathcal{F}f)(y)\overset{\vee}{\mathcal{F}}g(y)dy \\
&= \left\langle \tau_{-x}(\mathcal{F}f), \overline{\overset{\vee}{\mathcal{F}g}} \right\rangle \\
&= \langle \mathcal{F}(\chi_{-x}f), \overline{\mathcal{F}g} \rangle \\
&= \langle \chi_{-x}f, \overline{g} \rangle \\
&= \mathcal{F}(fg)(x).
\end{aligned}
$$
□

Remark 8.3.

1) The previous result is also valid for the Fourier cotransformation $\overline{\mathcal{F}}$.
2) In the same way, one shows that

$$\mathcal{F}(f * g) = \mathcal{F}f\mathcal{F}g \text{ and } \overline{\mathcal{F}}(f * g) = \overline{\mathcal{F}}f\overline{\mathcal{F}}g.$$

The link between the Fourier transformations in $\mathcal{S}'(\mathbb{R}^n)$ and in its subspace $L^2(\mathbb{R}^n)$ is given by the following.

Proposition 8.5. *The Fourier transformation in $\mathcal{S}'(\mathbb{R}^n)$ extends the one in $L^2(\mathbb{R}^n)$, i.e., $\hat{T}_f = T_{\hat{f}}$, for every $f \in L^2(\mathbb{R}^n)$.*

Proof. Let $f \in L^2(\mathbb{R}^n)$. For $\varphi \in \mathcal{D}(\mathbb{R}^n)$, one has

$$\left\langle \overset{\wedge}{T_f}, \varphi \right\rangle = \left\langle T_f, \overset{\wedge}{\varphi} \right\rangle = \int_{\mathbb{R}^n} f(x) \overset{\wedge}{\varphi}(x) dx = \int_{\mathbb{R}^n} \hat{f}(x) \varphi(x) dx.$$

To conclude use Parseval-Plancherel formula which expresses the preservation of the scalar product by the Fourier transformation $\mathcal{F}$. $\square$

8.2 Fourier range of $L^1(\mathbb{R}^n)$

The ordinary product of two functions in $L^2(\mathbb{R}^n)$ is a function in $L^1(\mathbb{R}^n)$. Conversely, any $f \in L^1(\mathbb{R}^n)$ can be written $f = gh$ with $g, h \in L^2(\mathbb{R}^n)$. Take for example

$$g(x) = \sqrt{|f(x)|} \text{ and } h(x) = \begin{cases} \frac{f(x)}{\sqrt{|f(x)|}} & \text{if } f(x) \neq 0, \\ 0 & \text{if } f(x) = 0. \end{cases}$$

Using Propositions 8.2 and 8.4 one obtains the following result:

Proposition 8.6. *The Fourier range of* $L^1(\mathbb{R}^n)$ *is exactly the set* $\{f * g : f, g \in L^2(\mathbb{R}^n)\}$.

Remark 8.4. Since $\mathcal{F}$ is a one-to-one map from $\mathcal{S}(\mathbb{R}^n)$ into $\mathcal{S}(\mathbb{R}^n)$, one has

$$\mathcal{S}(\mathbb{R}^n) \subset \mathcal{F}(L^1(\mathbb{R}^n)) \subset \mathcal{C}_0(\mathbb{R}^n),$$

where $\mathcal{C}_0(\mathbb{R}^n)$ is the space of continuous functions on $\mathbb{R}^n$ which vanish at infinity.

8.3 Fourier range of $\mathcal{O}_M(\mathbb{R}^n)$

We will characterize the Fourier range of $\mathcal{O}_M(\mathbb{R}^n)$, i.e., $\{\mathcal{F}T_f, f \in \mathcal{O}_M(\mathbb{R}^n)\}$. According to Corollary 6.1 of Chapter 6, this is reduced to characterize convolutors. Now any distribution with compact support is a convolutor by Proposition 6.19 of Chapter 6. On the other hand, by Proposition 7.2 of Chapter 7, there is a finite number of continuous functions f_β, $\beta \in \mathbb{N}^n$, having their supports in a neighborhood of *supp* T and such that $f = \sum_{\beta \in \mathbb{N}^n} D^\beta T_{f_\beta}$. A convolutor is a tempered distribution but it is not necessarily of compact support. So, we are led to examine the link between convolutors and tempered distributions of the form $\sum D^\beta T_{f_\beta}$, where the f_β's are continuous functions without any condition on their supports.

Proposition 8.7. *For a distribution $T \in \mathcal{S}'(\mathbb{R}^n)$, the following assertions are equivalent.*

1) $T = \mathcal{F}T_f$, with $f \in \mathcal{O}_M(\mathbb{R}^n)$.
2) *For every $m \in \mathbb{N}$, there is an integer $k_m \geq 0$ and a finite family of continuous functions (f_β), $|\beta| \leq k_m$, such that*
(i) $T = \sum\limits_{|\beta| \leq k_m} D^\beta T_{f_\beta}.$
(ii) *For any β, $|\beta| \leq k_m$, the function $x \longmapsto \left(1 + \|x\|^2\right)^m f_\beta$ is bounded on $\mathbb{R}^n$.*

Proof. 1) $\Longrightarrow$ 2) Let $f \in \mathcal{O}_M(\mathbb{R}^n)$. According to Proposition 6.9 of Chapter 6, f is of class $\mathcal{C}^\infty$ and slowly increasing. Hence, for every $m \in \mathbb{N}$, there is an $l \in \mathbb{N}$ depending on m such that, for any $\alpha \in \mathbb{N}^n, |\alpha| \leq 2m$, the function

$$x \longmapsto \frac{D^\alpha f(x)}{\left(1 + \|x\|^2\right)^l}$$

is bounded on $\mathbb{R}^n$. Replacing l by $l + n$ if necessary, we can suppose that this function is integrable on $\mathbb{R}$. The function g defined on $\mathbb{R}$ by

$$g(x) = \frac{f(x)}{\left(1 + \|x\|^2\right)^l}$$

is of class $\mathcal{C}^\infty$ and $D^\alpha g \in L^1(\mathbb{R}^n)$, for any $|\alpha| \leq 2m$. One then gets, according to (ii) of Theorem 5.2 in Chapter 5, that $M^\alpha \hat{g}$ is a continuous and bounded function on $\mathbb{R}^n$. Then for every $\varphi \in \mathcal{S}(\mathbb{R}^n)$,

$$\langle \mathcal{F}T_f, \varphi \rangle = \left\langle T_f, \overset{\wedge}{\varphi} \right\rangle$$
$$= \int_{\mathbb{R}^n} \left(1 + \|x\|^2\right)^l g(x) \overset{\wedge}{\varphi}(x) dx$$
$$= \sum \frac{l!}{p_0! \dots p_n!} \int_{\mathbb{R}^n} g(x) x_1^{2p_1} \dots x_n^{2p_n} \overset{\wedge}{\varphi}(x) dx,$$

where the summation extends over all finite sequences $p_0, p_1, \dots, p_n$ of natural integers such that $p_0 + \dots + p_n = l$. By the Exchange Formula (ii, Theorem 5.2 of Chapter 5), one has

$$x_1^{2p_1} \dots x_n^{2p_n} \overset{\wedge}{\varphi}(x) = \frac{1}{(-4\pi^2)^{p_1 + \dots + p_n}} D^{(2p_1, \dots, 2p_n)} \varphi(x).$$

Then, by the Transfer Formula (Theorem 5.3 of Chapter 5)

$$\langle \mathcal{F}T_f, \varphi \rangle = \sum \frac{l!}{p_0! \dots p_n!} \int_{\mathbb{R}^n} \frac{1}{(-4\pi^2)^{p_1 + \dots + p_n}} \hat{g}(x) D^{(2p_1, \dots, 2p_n)} \varphi(x) dx.$$

Whence

$$\mathcal{F}T_f = \sum c_{(p_0 \dots p_n)} . D^{(2p_1, \dots, 2p_n)} T_{\hat{g}}$$

where $c_{(p_0 \dots p_n)}$ is a positive constant. So, there is an integer $k_m \geq 0$ and a finite family of continuous functions $(f_\beta)_{|\beta| \leq k_m}$ of the form $c_\beta \hat{g}$ where the c_β are constants, such that

$$\mathcal{F}T_f = \sum_{|\beta| \leq k_m} D^\beta T_{f_\beta},$$

whence (i). On the other hand, for any β, $|\beta| \leq k_m$, one has, using Lemma 6.1 of Chapter 6

$$\left| \left(1 + \|x\|^2 \right)^m f_\beta(x) \right| \leq c_\beta \sup_{|\alpha| \leq m} \left| x^{2\alpha} \right| \left| \hat{g}(x) \right| \leq c \, c_\beta \sup_{|\alpha| \leq 2m} \left| x^\alpha \hat{g}(x) \right|,$$

whence (ii).

2) $\Longrightarrow$ 1) Let $T \in \mathcal{S}'(\mathbb{R}^n)$ satisfying (i) and (ii) of 2). Replacing n by $m + n$ if necessary, we can suppose that

$$x \longmapsto \left(1 + \|x\|^2 \right)^m f_\beta$$

is in $L^1(\mathbb{R}^n)$. Then, according to (i) of Lemma 6.1 in Chapter 6, $M^\alpha f_\beta \in L^1(\mathbb{R}^n)$ for any α such that $|\alpha| \leq 2m$. Hence $\overline{\mathcal{F}} f_\beta$ is of class $\mathcal{C}^{2m}$ by (i) of Theorem 5.2 in Chapter 5. Now

$$\begin{aligned}
\overline{\mathcal{F}}T &= \sum \overline{\mathcal{F}} \left(D^\beta T_{f_\beta} \right) \\
&= \sum (-2i\pi M)^\beta \, \overline{\mathcal{F}} T_{f_\beta} \\
&= \sum (-2i\pi M)^\beta \, T_{\overline{\mathcal{F}} f_\beta}, \quad \text{for } f_\beta \in L^1(\mathbb{R}^n) \\
&= T_{\sum (-2i\pi M)^\beta \overline{\mathcal{F}} f_\beta}.
\end{aligned}$$

So $\overline{\mathcal{F}}T$ is a regular distribution defined by a function g of class $\mathcal{C}^{2m}$, and this for every m. Hence it is defined by a function of class $\mathcal{C}^\infty$, due to the Du Bois-Reymond Lemma (cf. [15]). According to Proposition 6.9 of Chapter 6, it remains to show that $D^\alpha g$ is slowly increasing for every $\alpha \in \mathbb{N}^n$. Let m be such that $|\alpha| \leq 2m$. One has

$$D^\alpha g = \sum D^\alpha \left[(-2i\pi M)^\beta \, \overline{\mathcal{F}} f_\beta \right].$$

But

$$D^\alpha \left[(-2i\pi M)^\beta \overline{\mathcal{F}} f_\beta \right] = (-2i\pi)^\beta D^\alpha (M^\beta \overline{\mathcal{F}} f_\beta)$$

$$= (-2i\pi)^\beta \sum_{\gamma \leq \alpha} \frac{\alpha!}{\gamma!(\alpha-\gamma)!} D^\gamma M^\beta D^{\alpha-\gamma} \overline{\mathcal{F}} f_\beta$$

$$= (-2i\pi)^\beta \sum_{\gamma \leq \alpha} \frac{\alpha!}{\gamma!(\alpha-\gamma)!} (D^\gamma M^\beta) \overline{\mathcal{F}} \left[(2i\pi M)^{\alpha-\gamma} f_\beta \right].$$

Hence

$$\left| D^\alpha \left[(-2i\pi M)^\beta \overline{\mathcal{F}} f_\beta \right] \right| \leq c_1 \sum_{\gamma \leq \alpha} \frac{\alpha!}{\gamma!(\alpha-\gamma)!} \left| D^\gamma M^\beta \right|$$

$$\leq c_2 \sum_{\gamma \leq \alpha} \left| M^{\beta-\gamma} \right|$$

$$\leq c_3 \sup_{|\gamma| \leq \beta} \left| M^\gamma \right|$$

$$\leq c_4 \left(1 + \|x\|^2 \right)^{\frac{k}{2}},$$

where c_1, c_2, c_3 and c_4 are positive constants. So $D^\alpha g$ is slowly increasing. The proof is finished by $\mathcal{F}\overline{\mathcal{F}}T = T$. $\qquad\qquad\square$

8.4   Fourier range of $\mathcal{D}(\mathbb{R}^n)$

Let $\varphi \in \mathcal{D}(\mathbb{R}^n)$ and $a \geq 0$ such that $supp\, \varphi \subset \overline{B}(0,a)$. One has by definition

$$\overset{\wedge}{\varphi}(y) = \int_{\mathbb{R}^n} \varphi(x) e^{-2i\pi xy} dx; \quad y \in \mathbb{R}^n.$$

For every $z \in \mathbb{C}^n$, the function $x \longmapsto \varphi(x) e^{-2i\pi xz}$, where $xz = x_1 z_1 + \cdots + x_n z_n$, is integrable on $\mathbb{R}^n$, hence we put

$$\widetilde{\varphi}(z) = \int_{\mathbb{R}^n} \varphi(x) e^{-2i\pi xz} dx; \quad z \in \mathbb{C}^n.$$

The function $\widetilde{\varphi}$ obviously extends $\overset{\wedge}{\varphi}$. Moreover, it is a holomorphic function on $\mathbb{C}^n$ for it is differentiable with respect to each coordinate z_i of z, by Lebesgue's derivation theorem under the integral sign. And we have for $\alpha \in \mathbb{N}^n$

$$D^\alpha \widetilde{\varphi}(z) = \int_{\mathbb{R}^n} (-2i\pi x)^\alpha \varphi(x) e^{-2i\pi xz} dx.$$

As for the ordinary Fourier transform, one obtains

$$(2i\pi z)^\alpha \widetilde{\varphi}(z) = \int_{\mathbb{R}^n} (D^\alpha \varphi)(x) e^{-2i\pi xz} dx.$$

Writing $z = Re(z) + i \, Im(z)$ with $Re(z) = (Re(z_1), ..., Re(z_n))$ and $Im(z) = (Im(z_1), ..., Im(z_n))$, one has

$$|(2\pi z)^\alpha \widetilde{\varphi}(z)| \leq \int_{\|x\| \leq a} \left| (D^\alpha \varphi)(x) e^{-2i\pi x(Re(z) + iIm(z))} \right| dx$$

$$\leq \left(\sup_{\|x\| \leq a} e^{2\pi x Im(z)} \right) \|D^\alpha \varphi\|_1$$

$$\leq \|D^\alpha \varphi\|_1 \, e^{2\pi a \|Im(z)\|}.$$

But we know that, for every $k \in N$, there is a constant c depending on k such that

$$\left(1 + \|z\|^2 \right)^k |\widetilde{\varphi}(z)| \leq c \|D^\alpha \varphi\|_1 \, e^{2\pi a \|Im(z)\|}.$$

This inequality called Paley-Wiener and denoted (P.W), is essential in the characterization of the Fourier transforms of the elements of $\mathcal{D}(\mathbb{R}^n)$ as the following result shows.

Proposition 8.8 (Paley-Wiener theorem). *Let $a \geq 0$ and a function $f : \mathbb{R}^n \longrightarrow \mathbb{C}$. The following assertions are equivalent.*

(i) $f = \mathcal{F}\varphi$ with $\varphi \in \mathcal{D}(\mathbb{R}^n)$ such that $supp \, \varphi \subset \overline{B}(0, a)$.
(ii) f extends to a holomorphic function $\widetilde{f}$ on $\mathbb{C}^n$ satisfying the following property: For every $k \in \mathbb{N}$, there is $c_k \geq 0$ such that

$$(P.W) \quad \left(1 + \|z\|^2 \right)^k \left| \widetilde{f}(z) \right| \leq c_k e^{2\pi a \|Im(z)\|}; \ z \in \mathbb{C}^n.$$

Proof.   It remains to show that $(ii) \implies (i)$. So let $f : \mathbb{R}^n \longrightarrow \mathbb{C}$ be a function satisfying conditions of (ii). We look for $\varphi \in \mathcal{D}(\mathbb{R}^n)$ such that $supp \, \varphi \subset \overline{B}(0, a)$ and $f = \mathcal{F}\varphi$. The function f is in $L^1(\mathbb{R}^n)$ for by (P.W)

$$\left(1 + \|x\|^2 \right)^n |f(x)| \leq c_n, \text{ for every } x \in \mathbb{R}^n.$$

Let then $\varphi = \overline{\mathcal{F}} f$ i.e., $\varphi(x) = \int_{\mathbb{R}^n} f(y) e^{2i\pi xy} dy$. As by (P.W), the function $x \longmapsto \left(1 + \|x\|^2 \right)^k f(x)$ is in $L^1(\mathbb{R}^n)$ for every $k \in \mathbb{N}$, the function φ is of class $\mathcal{C}^\infty$ by Lemma 6.1 of Chapter 6 and (i) of Theorem 5.2 in Chapter 5.

On the other hand, $f \in L^2(\mathbb{R}^n)$ by (P.W). Hence $\mathcal{F}\varphi = \mathcal{F}(\overline{\mathcal{F}}f) = f$. So it remains only to show that $supp\ \varphi \subset \overline{B}(0, a)$. By Fubini's theorem, one has

$$\varphi(x) = \int_{\mathbb{R}} \cdots \left(\int_{\mathbb{R}} e^{2i\pi x_1 y_1} f(y_1, \ldots, y_n) dy_1 \right) e^{2i\pi x_2 y_2} \ldots e^{2i\pi x_n y_n} dy_2 \cdots dy_n.$$

Let $u \in \mathbb{R}^n$. Arguing as when calculating the Fourier transform of the function $x \longmapsto e^{-\pi x^2}$ (cf. Example 5.1 in Chapter 5) and taking into account the inequality

$$\left| \widetilde{f}(z) \right| \leq c_1 \left(1 + \|z\|^2 \right)^{-1} e^{2\pi a \|Im(z)\|},$$

one obtains

$$\int_{\mathbb{R}} e^{2i\pi x_1 (y_1 + iu_1)} f(y_1 + iu_1, y_2, \ldots, y_n) dy_1 = \int_{\mathbb{R}} e^{2i\pi x_1 y_1} f(y_1, y_2, \ldots, y_n) dy_1.$$

Repeating this argument for the other y_k one sees that, for every $x \in \mathbb{R}^n$ and every $u \in \mathbb{R}^n$,

$$\varphi(x) = \int_{\mathbb{R}^n} f(y + iu) e^{2i\pi x(y+iu)} dy.$$

Thus, we have

$$|\varphi(x)| \leq c_n\ e^{2\pi a \|u\| - 2\pi x u} \int_{\mathbb{R}^n} \frac{1}{\left(1 + \|y\|^2 \right)^n} dy$$

$$\leq c_n\ \pi^n\ e^{2\pi a \|u\| - 2\pi x u}.$$

Putting $u = \lambda x$ with $\lambda > 0$, one gets

$$|\varphi(x)| \leq c_n\ \pi^n\ e^{2\pi \lambda \|x\|(a - \|x\|)}$$

and this for any $\lambda > 0$. Letting λ tend to $+\infty$, one obtains $\varphi(x) = 0$ for $\|x\| > a$. $\qquad \square$

8.5 Fourier range of $\mathcal{E}'(\mathbb{R}^n)$

Let $T \in \mathcal{E}'(\mathbb{R}^n)$ and $a \geq 0$ such that $supp\ T \subset K$ where $K = \overline{B}(0, a)$. We know, by Proposition 6.23 of Chapter 6, that $\mathcal{F}T = T_f$ with f the function of class $\mathcal{C}^\infty$ given by

$$f(y) = \left\langle T_x, e^{-2i\pi xy} \right\rangle; \ y \in \mathbb{R}^n.$$

For every $z \in \mathbb{C}^n$, the function $x \longmapsto e^{-2i\pi xz}$ is still of class $\mathcal{C}^\infty$ on $\mathbb{R}^n$. Hence we can put

$$\widetilde{f}(z) = \left\langle T_x, e^{-2i\pi xz} \right\rangle; \ z \in \mathbb{C}^n.$$

The function $\widetilde{f}$ extends f. Moreover it is a holomorphic function on $\mathbb{C}^n$. As $supp\, T \subset K$ there is $m \in \mathbb{N}$ and $c > 0$ such that, for every $\varphi \in \mathcal{E}\left(\mathbb{R}^n\right)$

$$|\langle T, \varphi \rangle| \leq c \max_{\substack{|\alpha| \leq m \\ \|x\| \leq a}} \sup \left|(e^{-2i\pi z})^{\alpha} e^{-2i\pi xz}\right| \leq c(2\pi)^m \max_{|\alpha| \leq m} |z^{\alpha}| e^{2\pi a \|Im(z)\|}.$$

But

$$\sup_{|\alpha| \leq m} |z^{\alpha}| \leq \left(1 + \|z\|^2\right)^{\frac{m}{2}}.$$

Hence one has

$$\left(1 + \|z\|^2\right)^{-\frac{m}{2}} \left|\widetilde{f}(z)\right| \leq c_m \, e^{2\pi a \|Im(z)\|}$$

where c_m is a constant depending on m. This inequality called Paley-Wiener-Schwartz and denoted by (P.W.S.), is essential in the characterization of the Fourier transforms of distributions with compact supports, as it is shown by the following result.

Proposition 8.9 (Paley-Wiener-Schwartz theorem). *Let $\alpha \geq 0$ and $f : \mathbb{R}^n \longrightarrow \mathbb{C}$ a function. The following assertions are equivalent.*

(i) $T_f = \mathcal{F}T$ with $T \in \mathcal{E}'\left(\mathbb{R}^n\right)$ such that $supp\, T \subset \overline{B}\left(0, a\right)$.
(ii) f extends to a holomorphic function $\widetilde{f}$ on $\mathbb{C}^n$ satisfying the following property: There is $m \in \mathbb{N}$ and $c > 0$ such that

$$(P.\ W.\ S.) \quad \left|\widetilde{f}(z)\right| \leq c \left(1 + \|z\|^2\right)^{\frac{m}{2}} e^{2\pi a \|Im(z)\|}, \ z \in \mathbb{C}^n.$$

Proof. It remains to show that $(ii) \implies (i)$. Let $f : \mathbb{R}^n \longrightarrow \mathbb{C}$ be a function satisfying condition (ii). By (P.W.S.), the function f defines a slowly increasing measure. So there is $T \in \mathcal{S}'\left(\mathbb{R}^n\right)$ such that $T_f = \mathcal{F}T$ since $T_f \in \mathcal{S}'\left(\mathbb{R}^n\right)$. What remains to be shown is that $supp\, T \subset \overline{B}\left(0, a\right)$. Let $(\theta_j)_j$ be a regularizing sequence such that $supp\, \theta_j \subset \overline{B}(0, \varepsilon_j)$ where $(\varepsilon_j)_j$ tends to 0. Put $T_j = \theta_j * T$. Then

$$\mathcal{F}T_j = \overset{\wedge}{\theta}_j \mathcal{F}T = \overset{\wedge}{\theta}_j T_f = T_{\overset{\wedge}{\theta}_j f}.$$

According to the Paley-Wiener Theorem (Proposition 8.8), each $\overset{\wedge}{\theta}_j$ extends to a holomorphic function on $\mathbb{C}^n$, still denoted by $\overset{\wedge}{\theta}_j$. Moreover for every $p \in \mathbb{N}$, there exists a constant $c_{p,j}$ such that

$$\left(1 + \|z\|^2\right)^{\frac{p}{2}} \left|\overset{\wedge}{\theta}_j(z)\right| \leq c_{p,j} \, e^{2\pi \varepsilon_j \|Im(z)\|}.$$

Hence for every $k \in \mathbb{N}$, there is $c_{k,j} \geq 0$ such that

$$\left(1+\|z\|^{2}\right)^{\frac{k}{2}-\frac{m}{2}}\left|\overset{\wedge}{\theta_{j}}(z)\widetilde{f}(z)\right| \leq c_{k,j}\ e^{2\pi(\varepsilon_{j}+a)\|Im(z)\|}.$$

Applying this relation with $k' = k+m$, one sees that for every $k \in \mathbb{N}$, there exists $c'_{k,j} \geq 0$ such that

$$\left(1+\|z\|^{2}\right)^{\frac{k}{2}}\left|\overset{\wedge}{\theta_{j}}\widetilde{f}(z)\right| \leq c'_{k,j}\ e^{2\pi(\varepsilon_{j}+a)\|Im(z)\|}.$$

According to the Paley-Wiener Theorem again, one has $\overset{\wedge}{\theta_{j}}f = \mathcal{F}\varphi$ with $\varphi \in \mathcal{D}\left(\mathbb{R}^{n}\right)$ and $supp\ \varphi \subset \overline{B}(0, a+\varepsilon_{j})$. But

$$\mathcal{F}T_{j} = T_{\overset{\wedge}{\theta_{j}f}} = T_{\mathcal{F}\varphi} = \mathcal{F}T_{\varphi}.$$

And we have $T_{j} = T_{\varphi}$ for $\mathcal{F}$ is one-to-one, hence $supp\ T_{j} \subset \overline{B}(0, a+\varepsilon_{j})$. On the other hand, $\lim_{j} T_{j} = T$ in $\mathcal{D}'\left(\mathbb{R}^{n}\right)$. Let $\varphi \in \mathcal{D}\left(\mathbb{R}^{n}\right)$ be such that $(supp\ \varphi) \cap \overline{B}(0, a) = \emptyset$. For j large enough, one has $(supp\ \varphi) \cap \overline{B}(0, a+\varepsilon_{j}) = \emptyset$. Hence $\langle T_{j}, \varphi \rangle = 0$, whence $\langle T, \varphi \rangle = 0$.

$\square$

The following two consequences are worth mentioning:

Corollary 8.1. *The Fourier transform of a nonzero distribution with compact support has never a compact support, i.e.*

$$\mathcal{E}'\left(\mathbb{R}^{n}\right) \cap \mathcal{F}\left(\mathcal{E}'\left(\mathbb{R}^{n}\right)\right) = \{0\}.$$

Proof. This follows from the fact that an analytic function on $\mathbb{R}^{n}$ which has compact support is identically zero. $\square$

Corollary 8.2. *Let $T \in \mathcal{D}'\left(\mathbb{R}^{n}\right)$, such that $supp\ T = \{0\}$. Then there is $m \in \mathbb{N}$ and complex constants $c_{\alpha}, |\alpha| \leq m$, such that*

$$T = \sum_{|\alpha| \leq m} c_{\alpha}\ D^{\alpha}\delta.$$

Proof. We know that $\mathcal{F}T = T_{f}$ with $f(y) = \langle T_{x}, e^{-2i\pi xy} \rangle$, $y \in \mathbb{R}^{n}$. According to Proposition 8.9, the function f extends to $\mathbb{C}^{n}$ as a holomorphic function $\widetilde{f}$. Moreover there is $m \in \mathbb{N}$ and $c > 0$ such that

$$\left|\widetilde{f}(z)\right| \leq c \left(1+\|z\|^{2}\right)^{\frac{m}{2}}, \quad z \in \mathbb{C}^{n}.$$

As $\left(1 + \|z\|^2\right)^{\frac{m}{2}}$ is a polynomial, the Liouville theorem shows that $\widetilde{f}$ is still a polynomial. Hence there are complex constants a_α, $|\alpha| \leq m$, such that $\widetilde{f}(z) = \sum_{|\alpha| \leq m} a_\alpha z^\alpha$, for every $z \in \mathbb{C}^n$. One then has

$$T = \overline{\mathcal{F}} T_f$$
$$= \sum_{|\alpha| \leq m} a_\alpha \overline{\mathcal{F}}(T_{M^\alpha})$$
$$= \sum_{|\alpha| \leq m} a_\alpha \frac{1}{(2\pi)^\alpha} D^\alpha \delta$$
$$= \sum_{|\alpha| \leq m} c_\alpha D^\alpha \delta T. \qquad \square$$

It is well known that the Fourier transformation is among the most fundamental tools in different mathematical disciplines and also in physics. We have seen (Chap. 6, Section 6.7) that it can be defined even in $\mathcal{D}'(\mathbb{R}^n)$. By definition, $\mathcal{F}(\mathcal{D}'(\mathbb{R}^n))$ is exactly the space $\mathcal{Z}'(\mathbb{R}^n)$ of ultradistributions. In this course, the largest subspace of $\mathcal{D}'(\mathbb{R}^n)$ is $\mathcal{S}'(\mathbb{R}^n)$. It is then natural to characterize the Fourier ranges of some of its particular subspaces made of functions or distributions.

By construction, $\mathcal{S}(\mathbb{R}^n)$ is stable by the Fourier transform $\mathcal{F}$. Using the transfer formula one obtains that $\mathcal{F}$ is an isometry from $\mathcal{S}(\mathbb{R}^n)$ onto itself, for the norm of $L^2(\mathbb{R}^n)$. This is the Plancherel-Parseval theorem. Then $\mathcal{F}$ extends to an isometry of $L^2(\mathbb{R}^n)$ onto itself, due to the density of $\mathcal{S}(\mathbb{R}^n)$ in $L^2(\mathbb{R}^n)$. This is the Riesz-Plancherel theorem. Thus briefly

$$\mathcal{F}(\mathcal{S}(\mathbb{R}^n)) = \mathcal{S}(\mathbb{R}^n) \quad \text{and} \quad \mathcal{F}(L^2(\mathbb{R}^n)) = L^2(\mathbb{R}^n).$$

As a direct consequence, of the preceding, one has

$$\mathcal{F}(L^1(\mathbb{R}^n)) = L^2(\mathbb{R}^n) * L^2(\mathbb{R}^n).$$

The image of $\mathcal{D}(\mathbb{R}^n)$ by $\mathcal{F}$ is described by the Paley-Wiener theorem. It is exactly the space of functions $f : \mathbb{R}^n \longrightarrow \mathbb{C}$, which extend to a holomorphic function $\widetilde{f}$ on $\mathbb{C}^n$, satisfying the Paley-Wiener inequality. Concerning the space $\mathcal{O}_M(\mathbb{R}^n)$ of multipliers, one obtains

$$\mathcal{F}(\mathcal{O}_M(\mathbb{R}^n)) = \mathcal{O}'_c(\mathbb{R}^n),$$

where $\mathcal{O}'_c(\mathbb{R}^n)$ is the space of convoluters. It is a subspace of $\mathcal{S}'(\mathbb{R}^n)$. The description of $\mathcal{F}(\mathcal{O}_M(\mathbb{R}^n))$ given in Proposition 8.7, determines the structure of the convoluters. The analogue of Paley-Wiener theorem for distributions characterizes the Fourier image $\mathcal{F}(\mathcal{E}'(\mathbb{R}^n))$ of distributions of compact support $\mathcal{E}'(\mathbb{R}^n)$ (Paley-Wiener-Schawrtz theorem).

The results can, more or less, be found in the references but not in the same chapter. The description of the ranges $\mathcal{F}(\mathcal{D}(\mathbb{R}^n))$, $\mathcal{F}(\mathcal{E}'(\mathbb{R}^n))$ and $\mathcal{F}(\mathcal{O}_M(\mathbb{R}^n))$ are given in a section of [15]. We followed the idea of collecting results of the same type. We found it useful for the reader to extend the content to other subspaces of $\mathcal{S}'(\mathbb{R}^n)$ and make of it a separate chapter. More in [6], [12], [14] and [15].

Chapter 9

Laplace Transformation

We begin by the study of the Laplace transformation of real valued functions. We are interested in the existence domain and in some computational properties (among others derivation and integration formulas). We also give the inversion formula of Laplace. The multiplicative and convolution products are changed into each other by the Laplace transformation $\mathcal{L}$. Holomorphy of the function $\mathcal{L}f$ is also examined. For distributions we limit ourselves to those of supports in $\mathbb{R}_+$. The existence of abscissa is discussed and the holomorphy of the function $\mathcal{L}T$ the Laplace transform of a distribution is also examined. To be noticed is an interesting relation between the Laplace and the Fourier transformations. We also give the exchange, uniqueness and inversion theorems. Finally we present simple applications of the Laplace transformation in the resolution of Partial Differential Equations.

9.1 Laplace transforms of functions of real variables

We have seen (Proposition 8.8 of Chapter 8) that if $\varphi \in \mathcal{D}(\mathbb{R})$, then its Fourier transform $\overset{\wedge}{\varphi}$ given by

$$\overset{\wedge}{\varphi}(y) = \int_{\mathbb{R}} \varphi(x)e^{-2i\pi xy}dx$$

extends to $\mathbb{C}$ as a holomorphic function, again denoted $\overset{\wedge}{\varphi}$, given by

$$\overset{\wedge}{\varphi}(z) = \int_{\mathbb{R}} \varphi(x)e^{-2i\pi xz}dx.$$

This extension is called the Fourier-Laplace transform of φ. It is denoted $(\mathcal{F}\mathcal{L})(\varphi)$. If $f \in \mathcal{L}^1_{loc}(\mathbb{R})$ the expression

$$\int_{\mathbb{R}} f(x)e^{-2i\pi xz}dx$$

does not always have a meaning (take for example $f(x) = e^{2i\pi xz}$). So we are led to study the integrability of the function

$$t \longmapsto f(t)e^{-pt} \text{where } p \in \mathbb{C}.$$

Definition 9.1. Let $f \in \mathcal{L}^1_{loc}(\mathbb{R})$. We call the two sided (or bilateral) Laplace transform of f, in case of existence, the complex function denoted $\mathcal{L}f$ and given by

$$\mathcal{L}f(p) = \int_{\mathbb{R}} f(t)e^{-pt}dt; \ p \in \mathbb{C}.$$

For convenience, we will say only the Laplace transform of f.

Let us first examine the existence of $\mathcal{L}f$. If we put $p = a+ib$, a and b in $\mathbb{R}$, one has $|f(t)e^{-pt}| = |f(t)| \, e^{-at}$. The problem arises at the neighborhood of $+\infty$ and of $-\infty$. Notice that if there is $a_0 \in \mathbb{R}$ such that the function $t \longmapsto f(t)e^{-(a_0+ib)t}$ is integrable in the neighborhood of $+\infty$, then it is also so for the function $t \longmapsto f(t)e^{-(a+ib)t}$, for any $a \geq a_0$. Put

$$a_1 = \inf \left\{ u \in \mathbb{R} \colon t \longmapsto f(t)e^{-ut} \text{ is integrable in the neighborhood of } +\infty \right\}$$

with the convention $\inf \emptyset = +\infty$. For any $a > a_1$, the function $t \longmapsto f(t)e^{-at}$ is integrable in the neighborhood of $+\infty$ and it is not for any $a < a_1$. The case $a = a_1$ is to be examined.

Considering the least upper bound, with the convention $\sup \emptyset = -\infty$, one obtains the existence of an a_2 such that, for any $a < a_2$, the function $t \longmapsto f(t)e^{-at}$ is integrable in the neighborhood of $-\infty$ and it is not for any $a > a_2$. Here also the case $a = a_2$ has to be examined.

We have the following result:

Proposition 9.1. *If $a_1 < a_2$, then $\mathcal{L}f$ is defined on the open strip in $\mathbb{C}$, bounded by the straight lines $x = a_1$ and $x = a_2$. Moreover $\mathcal{L}f$ is not defined outside this strip.*

Example 9.1.

1) If $f \in \mathcal{D}(\mathbb{R})$, then $a_1 = -\infty$ and $a_2 = +\infty$, so we recapture the fact that in this case $\mathcal{L}f$ is defined in the whole of $\mathbb{C}$.

2) If $f(t) = e^{t^2}$, one has $a_1 = +\infty$ and $a_2 = -\infty$. Hence f does not admit a Laplace transform. Therefore it does not admit a Fourier transform.

3) In the previous example one has

$$t^{-n}\,|f(t)| \xrightarrow[|t|\to+\infty]{} +\infty, \text{ for every } n.$$

Such a function is said to be rapidly increasing. If f is rapidly increasing, one has $-\infty \le a_2 < 0 < a_1 \le +\infty$. Hence f does not admit a Laplace transform nor a Fourier one.

4) If $f \in \mathcal{S}(\mathbb{R})$, it is known that $f \in \mathcal{L}^1(\mathbb{R})$ and so $a_1 \le 0$. In fact one has $a_1 < 0$, for otherwise

$$\int_0^{+\infty} |f(-t)|\,e^{ut}dt = +\infty, \text{ for any } u < 0.$$

But then by the monotone convergence theorem of Lebesgue, one gets $f \notin \mathcal{L}^1(\mathbb{R})$. It is shown in the same way that $a_2 > 0$. So $-\infty \le a_1 < 0 < a_2 \le +\infty$. This is a generalization of Example 1) where $a_1 = -\infty$ and $a_2 = +\infty$.

5) If $f(t) = 0$ for any $t < 0$, one has $a_2 = +\infty$ and $-\infty \le a_1$. If $f \in \mathcal{D}(\mathbb{R}_+)$ then $a_1 = -\infty$. If $f \in \mathcal{S}(\mathbb{R}^n)$ then $-\infty \le a_1 < 0$. And if f is rapidly increasing, then $0 < a_1$.

6) If for $t \ge t_0 \ge 0$, there exists $k \in \mathbb{R}$ and $c > 0$ such that $|f(t)| \le ce^{-kt}$, then $a_1 \le k$. If $f(t) = Y(t)e^{-kt}$ then $a_1 \le k < a_2 = +\infty$.

In the sequel we will write

$$f \sqsupset \mathcal{L}f \text{ or } f(t) \sqsupset \mathcal{L}f(p)$$

to mean that $\mathcal{L}f$ is the Laplace transform of f. Then writing

$$f(t) \sqsupset \mathcal{L}f(p) \text{ on }]a_1, a_2[$$

one means that $\mathcal{L}f$ is defined on the open strip in $\mathbb{C}$ bounded by the straight lines $x = a_1$ and $x = a_2$.

First of all, here are some immediate properties.

Proposition 9.2. *Let* $f, g \in \mathcal{L}^1_{loc}(\mathbb{R})$. *If*

$$f(t) \sqsupset \mathcal{L}f(p) \text{ on }]a_1, a_2[$$

and

$$g(t) \sqsupset \mathcal{L}g(p) \text{ on }]b_1, b_2[$$

then

(i) $f(t) + g(t) \sqsupset \mathcal{L}f(p) + \mathcal{L}g(p)$ on $]\max(a_1, b_1), \min(a_2, b_2)[$.

(ii) $\quad\quad f(\lambda t) \sqsupset \dfrac{1}{\lambda}\mathcal{L}f(\dfrac{p}{\lambda})$ on $]\lambda a_1, \lambda a_2[$, *for any* $\lambda > 0$.

(iii) $\quad f(t - \lambda) \sqsupset e^{-\lambda p}\mathcal{L}f(p)$ on $]a_1, a_2[$, *for any* $\lambda \in \mathbb{R}$.

(iv) $\quad e^{-\lambda t}f(t) \sqsupset \mathcal{L}f(p + \lambda)$ on $]a_1 - \operatorname{Re}\lambda, a_2 - \operatorname{Re}\lambda[$, *for any* $\lambda \in \mathcal{C}$.

If $f(t) \sqsupset \mathcal{L}f(p)$ *on* $]a_1, a_2[$, *then*

$$\mathcal{L}f(p) = \int_{\mathbb{R}} f(t)e^{-at}e^{-ibt}dt.$$

For a fixed $a_0 \in]a_1, a_2[$ *and* $p = a_0 + 2\pi ib$, *one has*

$$\mathcal{L}f(a_0 + 2\pi ib) = \mathcal{F}\left(f(t)e^{-a_0 t}\right)(b).$$

So the function $t \longmapsto f(t)e^{-a_0 t}$ *and its Fourier transform are in* $\mathcal{L}^1(\mathbb{R})$.

Using the inversion formula of Fourier, one obtains the following result. It is the inversion formula of Laplace.

Proposition 9.3. *Let* $f \in \mathcal{L}^1_{loc}(\mathbb{R})$. *If* $a_0 \in]a_1, a_2[$ *then*

$$f(t) = \frac{1}{2i\pi}\int_{a_0 - i\infty}^{a_0 + i\infty} \mathcal{L}f(p)e^{pt}dp; \quad t \in \mathbb{R}.$$

Concerning the multiplicative and the convolution products, one has the following:

Proposition 9.4. *Let* $f, g \in \mathcal{L}^1_{loc}(\mathbb{R})$ *be such that*

$$f(t) \sqsupset \mathcal{L}f(p) \text{ on }]a_1, a_2[\text{ and } g(t) \sqsupset \mathcal{L}g(q) \text{ on }]b_1, b_2[.$$

Then

(i) If $fg \in \mathcal{L}^1_{loc}(\mathbb{R})$, *one has*

$$f(t)g(t) \sqsupset \frac{1}{2\pi i}\int_{a_0 - i\infty}^{a_0 + i\infty} \mathcal{L}f(p)\mathcal{L}g(s - p)dp \text{ on }]a_0 + b_1, a_0 + b_2[,$$

where $a_0 \in]a_1, a_2[$.

(ii) If $f * g \in \mathcal{L}^1_{loc}(\mathbb{R})$, *one has*

$$f * g(t) \sqsupset \mathcal{L}f(p)\mathcal{L}g(p) \text{ on }]\max(a_1, b_1), \min(a_2, b_2)[.$$

Proof.

(i) Using the previous proposition, one obtains

$$\mathcal{L}(fg)(s) = \int_{\mathbb{R}} f(t)g(t)e^{-st}dt$$

$$= \int_{\mathbb{R}} \left(\frac{1}{2\pi i} \int_{a_0-i\infty}^{a_0+i\infty} \mathcal{L}f(p)e^{pt}dp \right) g(t)e^{-st}dt.$$

Since

$$\int_{\mathbb{R}} \left(\int_{a_0-i\infty}^{a_0+i\infty} |\mathcal{L}f(p)|\,|g(t)|\,e^{(a_0-\mathrm{Re}\,s)t}dp \right) dt < +\infty,$$

Fubini's theorem shows that $(t,p) \longmapsto |\mathcal{L}f(p)e^{pt}g(t)e^{-st}|$ is integrable. Hence we have

$$\mathcal{L}(fg)(s) = \frac{1}{2\pi i} \int_{a_0-i\infty}^{a_0+i\infty} \mathcal{L}f(p) \left(\int_{\mathbb{R}} g(t)e^{(p-s)t}dt \right) dp$$

$$= \frac{1}{2\pi i} \int_{a_0-i\infty}^{a_0+i\infty} \mathcal{L}f(p)\mathcal{L}g(s-p)dp.$$

(ii) One gets

$$\mathcal{L}(f*g)(s) = \int_{-\infty}^{+\infty} (f*g)(t)e^{-st}dt$$

$$= \int_{-\infty}^{+\infty} \left(\int_{-\infty}^{+\infty} f(x)g(t-x)dx \right) e^{-st}dt.$$

Since

$$\int_{-\infty}^{+\infty} \left(\int_{-\infty}^{+\infty} |f(x)g(t-x)e^{-st}|\,dt \right) dx < +\infty,$$

Fubini's theorem shows that $(x,t) \longmapsto |f(x)g(t-x)e^{-st}|$ is integrable. Hence we have

$$\mathcal{L}(f*g)(s) = \int_{-\infty}^{+\infty} \left(\int_{-\infty}^{+\infty} f(x)g(t-x)e^{-st}dt \right) dx$$

$$= \int_{\mathbb{R}} \left(\int_{\mathbb{R}} g(t-x)e^{-st}dt \right) f(x)dx$$

$$= \int_{\mathbb{R}} \left(\int_{\mathbb{R}} g(u)e^{-s(u+x)}du \right) f(x)dx,$$

whence the result.

$\square$

We now examine the holomorphy of the function $p \longmapsto \mathcal{L}f(p)$.

Proposition 9.5. *Let $f \in \mathcal{L}^1_{loc}(\mathbb{R})$ such that*

$$f(t) \sqsupset \mathcal{L}f(p) \ \text{on} \ ]a_1, a_2[.$$

Then the function $\mathcal{L}f$ is holomorphic on the open strip bounded by the straight lines $x = a_1$ and $x = a_2$. Moreover for every $m \in \mathbb{N}$,

$$(\mathcal{L}f)^{(m)}(p) = \int_{\mathbb{R}} (-t)^m f(t) e^{-pt} dt$$

i.e.,

$$(-t)^m f(t) \sqsupset (\mathcal{L}f)^{(m)}(p) \ \text{on} \ ]a_1, a_2[.$$

Proof. We first show that the domain of existence, determined by $]a'_1, a'_2[$, of the integral $\int_{-\infty}^{+\infty} (-t)^m f(t) e^{-pt} dt$ is the same as the one of $\mathcal{L}f$. We argue at the neighborhood of $+\infty$. It is sufficient to consider $t \geq 1$. In this case

$$\left| (-t)^m f(t) e^{-pt} \right| \leq \left| f(t) e^{-pt} \right|.$$

Whence $a_1 \leq a'_1$. Conversely, for any $\varepsilon > 0$ one has for t large enough,

$$\left| (-t)^m f(t) e^{-pt} \right| \leq \left| f(t) e^{-(p-\varepsilon)t} \right|$$

which gives $a'_1 \leq a_1 + \varepsilon$, for any ε. Hence $a'_1 \leq a_1$. We argue in the same way at the neighborhood of $-\infty$. To obtain the derivation formula, use Lebesgue's theorem on the derivation under the integral sign. $\qquad\square$

Here are other results:

Proposition 9.6. *Let $f \in \mathcal{L}^1_{loc}(\mathbb{R})$ such that*

$$f(t) \sqsupset \mathcal{L}f(p) \ \text{on} \ ]a_1, a_2[.$$

(i) If f is m times differentiable and $f^{(m)} \in \mathcal{L}^1_{loc}(\mathbb{R})$, then

$$f^{(m)}(t) \sqsupset p^m \mathcal{L}f(p) \ \text{on} \ ]a_1, a_2[.$$

(ii) If the function $t \longmapsto \int_{-\infty}^t f(u)du$ is differentiable with derivative $t \longmapsto f(t)$, then

$$\int_{-\infty}^t f(u)du \sqsupset \frac{\mathcal{L}f(p)}{p} \ \text{on} \ ]\max(0, a_1), a_2[.$$

Proof.

(i) If $m = 1$ one obtains, integrating by parts, that $\mathcal{L}f'(p) = p\mathcal{L}f(p)$ for $[f(t)e^{-pt}]_{-\infty}^{+\infty} = 0$, since $t \longmapsto f(t)e^{-pt}$ is in $\mathcal{L}^1(\mathbb{R})$. If $m \geq 2$ apply that result m times in a row.

(ii) This follows from (i). $\qquad\qquad\qquad\qquad\qquad\qquad\qquad\qquad\square$

Example 9.2.

1) $Y(t) \sqsupset \frac{1}{p}$ on $]0, +\infty[$.

2) $Y(t)\frac{t^m}{m!} \sqsupset \frac{1}{p^{m+1}}$ on $]0, +\infty[$.

3) $Y(t)\frac{t^m}{m!}e^{at} \sqsupset \frac{1}{(p-a)^{m+1}}$ on $]Re\, a, +\infty[$.

4) More generally, for $\alpha > 0$

$$Y(t)\frac{t^\alpha}{\Gamma(\alpha+1)}e^{at} \sqsupset \frac{1}{(p-a)^{\alpha+1}} \text{ on }]Re\, a, +\infty[.$$

Indeed, by (iv) of Proposition 9.2, it suffices to show that

$$Y(t)\frac{t^\alpha}{\Gamma(\alpha+1)} \sqsupset \frac{1}{p^{\alpha+1}} \text{ on }]0, +\infty[.$$

Let $p = re^{i\theta}$ with $\theta \in]-\frac{\pi}{2}, \frac{\pi}{2}[$. One has

$$\mathcal{L}\left(\frac{Y(t)t^\alpha}{\Gamma(\alpha+1)}\right) = \int_0^{+\infty} \frac{t^\alpha}{\Gamma(\alpha+1)}e^{-pt}dt$$

$$= \frac{1}{p^{\alpha+1}}\frac{1}{\Gamma(\alpha+1)}\int_L z^\alpha e^{-z}dz, \text{ where } z = tp,$$

L being the half-line of origin 0 determined by $\arg z = -\arg p$. Let $z = Re^{i\beta}$ and C_R be the arc of the circle centered at 0 and of radius R such that $-\alpha < \arg z < 0$. Then

$$\left|\int_{C_R} z^\alpha e^{-pz}dz\right| \leq R^\alpha \left|\int_{-\alpha}^0 e^{-rR\cos|\alpha+\beta|}Rd\beta\right|$$

$$\leq \alpha\, R^{\alpha+1}e^{-rR\cos\alpha} \text{ with } 0 < \alpha + \beta < \alpha.$$

Hence $\lim\limits_{R\to+\infty} \int_{C_R} z^\alpha e^{-pz}dz = 0$. By the Cauchy theorem, one has

$$\int_L z^\alpha e^{-z}dz = \int_0^{+\infty} t^\alpha e^{-t}dt.$$

5) From 4), one obtains that

$$Y(t)\cos\omega t \sqsupset \frac{1}{p^2+\omega^2} \text{ on }]0, +\infty[$$

and

$$Y(t)\sin\omega t \sqsupset \frac{\omega}{p^2+\omega^2} \text{ on }]0, +\infty[.$$

9.2   Laplace transforms of distributions

We limit ourselves to distributions with support in $\mathbb{R}_+$. This is what is more useful in practice. Let $f \in C(\mathbb{R})$ have compact support contained in $\mathbb{R}_+$. Then for $Re\ p > a_1$,

$$\mathcal{L}f(p) = \int_{\mathbb{R}} f(t)e^{-pt}dt = \left\langle T_f, e^{-pt} \right\rangle.$$

Let $f \in \mathcal{L}^1_{loc}(\mathbb{R})$ with $supp\ f \in \mathbb{R}_+$. One has

$$\mathcal{L}f(p) = \int_0^{+\infty} f(t)e^{-pt}dt \text{ on } Re\ p > a_1.$$

The latter integral is formally written $\langle T_f, e^{-pt} \rangle$ when it has a meaning. For $a_0 > a_1$, the function $g : t \longmapsto e^{-a_0 t} f(t)$ is in $\mathcal{L}^1(\mathbb{R})$ and hence defines a slowly increasing measure. So $T_g \in \mathcal{S}'(\mathbb{R})$. On the other hand, for every function ψ of class C^∞, equal to 1 on a neighborhood of $\mathbb{R}_+$ with support limited on the left, one gets that

$$\mathcal{L}f(p) = \int_{\mathbb{R}} e^{-a_0 t} f(t)\psi(t)e^{-(p-a_0)t}dt.$$

But $t \longmapsto \psi(t)e^{-(p-a_0)t}$ is in $\mathcal{S}(\mathbb{R})$, hence

$$\mathcal{L}f(p) = \left\langle e^{-a_0 t} f(t), \psi(t)e^{-(p-a_0)t} \right\rangle.$$

The latter expression does not depend of a_0 nor of ψ. Then we put

$$\left\langle T_f, e^{-pt} \right\rangle = \left\langle e^{-a_0 t} f(t), \psi(t)e^{-(p-a_0)t} \right\rangle.$$

Hence we can admit, by definition, that

$$\mathcal{L}T_f(p) = \left\langle T_f, e^{-pt} \right\rangle.$$

We will now treat a distribution T with support in $\mathbb{R}_+$, i.e., $T \in \mathcal{D}'(\mathbb{R}_+)$. Let $T \in \mathcal{D}'(\mathbb{R}_+)$. Put

$$\Gamma_T = \left\{ a \in \mathbb{R} : e^{-at}T \in \mathcal{S}'(\mathbb{R}) \right\}.$$

If Γ_T is non void, let $a \in \Gamma_T$ and $b > a$. For every ψ of class C^∞ on $\mathbb{R}$, equal to 1 on a a neighborhood of $\mathbb{R}_+$, one has

$$e^{-bt}T = \left(e^{-(b-a)t}\psi(t) \right)(e^{-at}T).$$

But $e^{-at}T \in \mathcal{S}'(\mathbb{R})$ and $t \longmapsto e^{-(b-a)t}\psi(t)$ is in $\mathcal{S}(\mathbb{R})$. Hence $e^{-bt}T \in \mathcal{S}'(\mathbb{R})$. So Γ_T is a half-line contained in $\mathbb{R}_+$. Put $\gamma_T = \inf \Gamma_T$. Then for

$\gamma_T < \gamma_0 < Re\ p$, the expression $\left\langle e^{-\gamma_0 t}T, \psi(t)e^{-(p-\gamma_0)t}\right\rangle$ has a meaning. It does not depend of γ_0 nor of ψ. Hence one can put, by definition,

$$\left\langle T, e^{-pt}\right\rangle = \left\langle e^{-\gamma_0 t}\ T, \psi(t)e^{-(p-\gamma_0)t}\right\rangle.$$

Definition 9.2. Let $T \in \mathcal{D}'(\mathbb{R}_+)$ such that $\Gamma_T \neq \emptyset$. We call the Laplace transform of T, the function denoted $\mathcal{L}(T)$ and given by

$$\mathcal{L}(T)(p) = \left\langle T, e^{-pt}\right\rangle, \text{ for } Re\ p > \gamma_T.$$

The open half-plane $\{p \in \mathbb{C}: Re\ p > \gamma_T\}$ is said to be the existence domain of $\mathcal{L}(T)$ and γ_T its existence abscissa.

Remark 9.1. If $T \in \mathcal{D}'(\mathbb{R}_+) \cap \mathcal{E}'(\mathbb{R})$, then $\Gamma_T = \mathbb{R}$. Hence for every $\gamma \in \mathbb{R}$ and every ψ of class $\mathcal{C}^\infty$ on $\mathbb{R}$, equal to 1 on $\mathbb{R}_+$, one has

$$\left\langle e^{-\gamma t}T, \psi(t)e^{-(p-\gamma)t}\right\rangle = \left\langle e^{-\gamma t}T, e^{-(p-\gamma)t}\right\rangle,$$

and as $e^{-\gamma t}T$ has compact support,

$$\left\langle e^{-\gamma t}T, e^{-(p-\gamma)t}\right\rangle = \left\langle T, e^{-pt}\right\rangle$$

for every $p \in \mathbb{C}$.

Remark 9.2. Let $T \in \mathcal{D}'(\mathbb{R}_+)$ with $\Gamma_T \neq \emptyset$. Then the function $\mathcal{L}T$ is holomorphic on the open half-plane $\overset{\circ}{\Gamma}_T \times \mathbb{R}$ and, for every $k \in \mathbb{N}$

$$(\mathcal{L}T)^{(k)} = ((-1)^k \mathcal{L}(M^k T) \text{ with } M^k(x) = x^k, \ x \in \mathbb{R}.$$

Proof. We argue by induction. It suffices to consider the case $k = 1$. One has

$$(\mathcal{L}T)(p) = \left\langle e^{-\gamma_0 t}T, \psi(t)e^{-(p-\gamma_0)t}\right\rangle \text{ with } \gamma_T < \gamma_0 < Re\ p.$$

And

$$\frac{(\mathcal{L}T)(p+h) - \mathcal{L}T(p)}{h} = \left\langle e^{-\gamma_0 t}T, \psi(t)\frac{e^{-(p+h+\gamma_0)t} - e^{-(p-\gamma_0)t}}{h}\right\rangle$$

$$= \left\langle e^{-\gamma_0 t}T, \psi(t)\frac{e^{-[(p-\gamma_0)+h]t} - e^{-(p-\gamma_0)t}}{h}\right\rangle.$$

This leads us to consider

$$\psi(t)\frac{e^{-(q+h)t} - e^{qt}}{h} \text{ with } Re\ q > 0.$$

When h tends to zero, this function tends to $t \longmapsto -t\psi(t)e^{-qt}$ in $\mathcal{S}(\mathbb{R})$, by the following lemma. $\qquad\square$

Lemma 9.1. *Let $p \in \mathbb{C}$ with Re $p > 0$, $h \in \mathbb{C}^*$ with $2|h| <$ Re p and $f \in \mathcal{O}_M(\mathbb{R})$ with supp f limited on the left. Put*

$$g_h(t) = f(t)\frac{e^{-(p+h)t} - e^{pt}}{h} \text{ and } g(t) = -f(t)te^{-pt}.$$

Then g_h tends to g in $\mathcal{S}(\mathbb{R})$ when h tends to 0.

Proof. We will show that

(i) g_h tends to g in $\mathcal{E}(\mathbb{R})$ when h tends to 0.
(ii) $\forall k \in \mathbb{N}, \forall m \in \mathbb{N}, \exists c_{k,m} > 0$ such that

$$\sup_h |g_h|_{k,m} \leq c_{k,m}.$$

which will imply the result. Indeed, replacing g_h by $g_h - g$ if necessary, we can suppose that g_h tends to 0. Let $k \in \mathbb{N}$ and $m \in \mathbb{N}$. By (ii) there is $c_{k+1,m} > 0$ such that

$$|g_h|_{k+1,m} \leq c_{k+1,m}, \ \forall h$$

i.e.,

$$\max_{|\alpha| \leq m} \sup_{t \in \mathbb{R}} \left|(1 + |t|^2)^{k+1} g_h^{(\alpha)}(t)\right| \leq c_{k+1,m}, \ \forall h.$$

Whence, for every $t \in \mathbb{R}$

$$\max_{|\alpha| \leq m} \left|(1 + |t|^2)^k g_h^{(\alpha)}(t)\right| \leq \frac{c_{k+1,m}}{1 + |t|^2}.$$

For $\varepsilon > 0$, there is $R > 0$ such that

$$\frac{c_{k+1,m}}{1 + |t|^2} < \frac{\varepsilon}{2} \text{ for } |t| > R.$$

And by (i)

$$\max_{|\alpha| \leq m} \sup_{t \in \mathbb{R}} \left|(1 + |t|^2)^k g_h^{(\alpha)}(t)\right| \leq \frac{\varepsilon}{2} \text{ for } |h| \text{ small enough.}$$

It remains to show (i) and (ii).

(i) One has

$$g_h(t) - g(t) = f(t)e^{-pt}\left(\frac{e^{-ht} - 1}{h} + t\right)$$

which tends to zero in $\mathcal{E}(\mathbb{R})$ when h tends to 0. This follows from the fact that $f \in \mathcal{O}_M(\mathbb{R})$ and the fact that $\dfrac{e^{-ht} - 1}{h} + t$ tends to 0 in $\mathcal{E}(\mathbb{R})$

when h tends to 0.

(ii) One gets

$$g_h(t) = f(t)e^{-pt}\frac{e^{-ht}-1}{h} \text{ and } \left|\frac{e^{-ht}-1}{h}\right| \le |t|\, e^{|ht|},$$

hence for every $k \in \mathbb{N}$

$$\left|(1+|t|^2)^k g_h(t)\right| \le \left|(1+|t|^2)^{k+1} f(t)e^{-pt+|ht|}\right|,$$

and

$$\sup_h \sup_{t\in\mathbb{R}} \left|(1+|t|^2)^k g_h(t)\right| < +\infty$$

in view of $|h| \le \frac{Re\, p}{2}$, i.e., $\sup_h |g_h|_{k,0} < +\infty$. Similarly, for every $m \in \mathbb{N}^*$, there is $c_m > 0$, such that

$$\left|\left(\frac{e^{-ht}-1}{h}\right)^{(m)}\right| \le c_m\, e^{|ht|}.$$

Whence

$$\sup_h \sup_{t\in\mathbb{R}} \left|(1+|t|^2)^k g_h^{(m)}(t)\right| < +\infty, \text{ i.e., } \sup_h |g_h|_{k,m} < +\infty. \qquad \square$$

Proposition 9.7 (Exchange theorem). *Let $S, T \in \mathcal{D}'(\mathbb{R}_+)$. Then*

(i) $\Gamma_S \cap \Gamma_T \subset \Gamma_{S*T}$.
(ii) $\mathcal{L}(S*T)(p) = \mathcal{L}(S)(p)\,\mathcal{L}(T)(p)$ *for Re $p > \max(\gamma_S, \gamma_T)$.*

Proof.

(i) If $\gamma \in \Gamma_S \cap \Gamma_T$ then, by Proposition 6.21 of Chapter 6,

$$(e^{-\gamma t}S) * (e^{-\gamma t}T) \in \mathcal{D}'(\mathbb{R}_+) \cap \mathcal{S}'(\mathbb{R}).$$

But

$$(e^{-\gamma t}S) * (e^{-\gamma t}T) = e^{-\gamma t}(S*T).$$

So $\gamma \in \Gamma_{S*T}$.

(ii) Let $p \in \mathbb{C}$ such that $Re\, p > \max(\gamma_S, \gamma_T)$ and let ψ be a function of class $\mathcal{C}^\infty$ on $\mathbb{R}$, with a support limited on the left and equal to 1 on a neighborhood of $\mathbb{R}_+$. For $\max(\gamma_S, \gamma_T) < \gamma_1 < Re\, p$, one has

$$\begin{aligned}\mathcal{L}(S*T)(p) &= \left\langle e^{-\gamma_1 t}(S*T), \psi(t)e^{-(p-\gamma_1)t}\right\rangle\\ &= \left\langle (e^{-\gamma_1 t}S)*(e^{-\gamma_1 t}T), \psi(t)(e^{-(p-\gamma_1)t})\right\rangle\\ &= \left\langle (e^{-\gamma_1 t}S)\otimes(e^{-\gamma_1 t}T), \psi(t)(e^{-(p-\gamma_1)t})\right\rangle.\end{aligned}$$

As $\psi^\Delta = \psi \otimes \psi$ on a neighborhood of $(supp\ S) \times (supp\ T)$ and

$$\psi \otimes \psi \left[e^{-(p-\gamma)t} \right]^\Delta = e^{-(p-\gamma_1)t} \psi \otimes e^{-(p-\gamma_1)t} \psi,$$

one has

$$\mathcal{L}(S * T)(p) = \left\langle e^{-\gamma_1 t} S, \psi(t) e^{-(p-\gamma_1)t} \right\rangle \left\langle e^{-\gamma_1 t} T, \psi(t) e^{-(p-\gamma_1)t} \right\rangle$$
$$= \mathcal{L}(S)(p)\mathcal{L}(T)(p).$$

$\square$

The link between Fourier and Laplace transformations is given the by following.

Proposition 9.8. *Let $T \in \mathcal{D}'(\mathbb{R}_+)$ and γ_T be the existence abscissa of $\mathcal{L}T$. Then for $\gamma > \gamma_T$, the Fourier transform of the tempered distribution $e^{-\gamma t} T$ is a regular distribution defined by the function f_γ given by*

$$f_\gamma(x) = \mathcal{L}(T)(\gamma + 2i\pi x), \quad x \in \mathbb{R}.$$

Proof. We will show that $\mathcal{F}(e^{-\gamma t} T) = T_{f_\gamma}$. For every $\varphi \in \mathcal{D}(\mathbb{R})$, one has

$$\langle T_{f_\gamma}, \varphi \rangle = \int_{\mathbb{R}} \mathcal{L}(T)(\gamma + 2i\pi x)\varphi(x)dx$$
$$= \int_{\mathbb{R}} \left\langle e^{-\gamma_1 t} T, \psi(t) e^{-(\gamma + 2i\pi x - \gamma_1)t} \right\rangle \varphi(x)dx$$
$$= \int_{\mathbb{R}} \left\langle e^{-\gamma_1 t} T, \psi(t) e^{-(\gamma - \gamma_1)t} \right\rangle e^{-2i\pi x t} \varphi(x)dx$$
$$= \int_{\mathbb{R}} \left\langle e^{-\gamma_1 t} T, \psi(t) e^{-(\gamma - \gamma_1)t} \right\rangle \left\langle (T_\varphi)_x, e^{-2i\pi x t} \right\rangle.$$

And using the series expansion of $e^{-2i\pi x t}$, one obtains

$$\langle T_{f_\gamma}, \varphi \rangle = \left\langle e^{-\gamma_1 t} T \otimes (T_\varphi)_x, \psi(t) e^{-(\gamma - \gamma_1)t} e^{-2i\pi x t} \right\rangle$$
$$= \left\langle e^{-\gamma t} T, \left\langle (T_\varphi)_x, e^{-2i\pi x t} \right\rangle \right\rangle$$
$$= \left\langle e^{-\gamma t} T, \mathcal{F}(\varphi(t)) \right\rangle$$
$$= \left\langle \mathcal{F}\left(e^{-\gamma t} T \right), \varphi \right\rangle.$$

$\square$

Corollary 9.1 (Uniqueness theorem). *Let $S, T \in \mathcal{D}'(\mathbb{R}_+)$ admitting Laplace transforms. If there exists $\gamma_0 \in \Gamma_S \cap \Gamma_T$ such that*

$$\mathcal{L}(S)(\gamma_0 + 2i\pi x) = \mathcal{L}(T)(\gamma_0 + 2i\pi x), \quad \text{for every } x \in \mathbb{R},$$

then $S = T$.

Proof.   We can suppose $\gamma_0 > \max(\gamma_S, \gamma_T)$. Then by the previous proposition,

$$\mathcal{F}(e^{-\gamma_0 t} S) = \mathcal{F}(e^{-\gamma_0 t} T).$$

Whence $e^{-\gamma_0 t} S = e^{-\gamma_0 t} T$. Hence $S = T$.

$\square$

If $S, T \in \mathcal{D}'(\mathbb{R}_+)$ admit Laplace transforms, then $\mathcal{L}(S)(p)\mathcal{L}(T)(p) = 1$ for $Re\ p > Max(\gamma_S, \gamma_T)$ whenever $S * T = \delta$, since $\mathcal{L}(\delta)(p) = 1$. The converse is also true.

Proposition 9.9 (Inversion theorem). *Let $S, T \in \mathcal{D}'(\mathbb{R}_+)$ admitting Laplace transforms. If*

$$\mathcal{L}(S)(p)\mathcal{L}(T)(p) = 1 \text{ for } Re\ p > Max(\gamma_S, \gamma_T),$$

then $S * T = \delta$, i.e., S and T are inverse of each other in the convolution algebra $\mathcal{D}'(\mathbb{R}_+)$.

Proof.   By the Exchange Theorem (Proposition 9.7), one has

$$\mathcal{L}(S * T)(p) = \mathcal{L}(S)(p)\mathcal{L}(T)(p) = 1.$$

Whence the result by Corollary 9.1.

$\square$

Example 9.3.

1) For $b \geq 0$, one has $\mathcal{L}(\delta_b) = e^{-bp}$; $p \in \mathbb{C}$.
2) For $m \in \mathbb{N}$, one has $\mathcal{L}(\delta^{(m)})(p) = p^m$; $p \in \mathbb{C}$.
3) If $f \in \mathcal{L}^1_{loc}(\mathbb{R})$ with $supp\ f \subset \mathbb{R}_+$, then $\mathcal{L}(T_f)(p)) = \mathcal{L}(f)(p)$ for $Re\ p > 0$. In particular $\mathcal{L}(T_Y)(p) = \frac{1}{p}$, for $Re\ p > 0$.

Recapitulative table 2.10

$m \in \mathbb{N}$, $a \in \mathbb{C}$, $\omega \in \mathbb{R}$ and $\beta \in \mathbb{R}$.

original	image	existence domain of the image
$Y(t)$	$\frac{1}{p}$	$Re\ p > 0$
$Y(t)\frac{t^m}{m!}$	$\frac{1}{p^{m+1}}$	$Re\ p > 0$
$Y(t)\frac{t^m}{m!}e^{at}$	$\frac{1}{(p-a)^{m+1}}$	$Re\ p > Re\ a$
$Y(t)\frac{t^\alpha}{\Gamma(\alpha+1)}e^{at}$, $\alpha > 0$	$\frac{1}{(p-a)^{\alpha+1}}$	$Re\ p > Re\ a$
$Y(t)\cos\omega t$	$\frac{p}{p^2+\omega^2}$	$Re\ p > 0$
$Y(t)\sin\omega t$	$\frac{\omega}{p^2+\omega^2}$	$Re\ p > 0$
$Y(t)e^{-\beta t}\cos\omega t$	$\frac{p+\beta}{(p+\beta)^2+\omega^2}$	$Re\ p > -\beta$
$Y(t)e^{-\beta t}\sin\omega t$	$\frac{\omega}{(p+\beta)^2+\omega^2}$	$Re\ p > -\beta$
δ	1	$\mathbb{C}$
δ_b, $b \geq 0$	e^{-bp}	$\mathbb{C}$
$\delta^{(m)}$	p^m	$\mathbb{C}$
$\delta_b^{(m)}$	$p^m e^{-bp}$	$\mathbb{C}$

Now here are simple examples of PDE's in the resolution of which the Laplace transform proves its efficiency.

Consider a convolution equation $A * X = B$, in a given convolution algebra. If A, X and B admit Laplace transforms, this equation is equivalent to $\mathcal{L}(A)\,\mathcal{L}(X) = \mathcal{L}(B)$. Under suitable conditions, according to the case in hand, the solution is given, when it exists, by

$$X = \mathcal{L}^{-1}\left[\frac{\mathcal{L}(B)}{\mathcal{L}(A)}\right].$$

In the following particular case

$$\left(\sum_{j=0}^{m}\alpha_j\delta^{(j)}\right) * X = B,$$

one has

$$\left(\sum_{j=0}^{m}\alpha_j\,p^j\right)\mathcal{L}X(p) = \mathcal{L}B(p).$$

Hence when the solution exists, it is given by

$$X = \mathcal{L}^{-1}\left(\frac{\mathcal{L}B(p)}{\sum_{j=0}^{m}\alpha_j p^{(j)}}\right).$$

Here are some examples as an application.

1) **Elementary solution of** $D + \lambda, \lambda \in \mathbb{C}$. We look for E in $\mathcal{D}'(\mathbb{R}_+)$ such that

$$(\delta' + \lambda\delta) * E = \delta.$$

Which gives

$$(p + \lambda)\mathcal{L}E(p) = 1.$$

So

$$\mathcal{L}E(p) = \frac{1}{p + \lambda}; \text{ for } Re\ p > -Re\ \lambda.$$

But one has,

$$Y(t)e^{-\lambda t} \sqsupset \frac{1}{p + \lambda}; \text{ for } Re\ p > -Re\ \lambda.$$

Hence

$$E = T_{Ye^{-\lambda t}}.$$

So we find again the result in 1), of Example 4.4 of Chapter 4.

2) **Elementary solution of** $(D + \lambda)^m$, $\lambda \in \mathbb{C}$ **with** $m \in N^*$. We look for E in $\mathcal{D}'(\mathbb{R}_+)$ such that

$$(D + \lambda)^m E = \delta,$$

which can be written

$$(\delta' + \lambda\delta)^m * E = \delta$$

where

$$(\delta' + \lambda\delta)^m = (\delta' + \lambda\delta) * (\delta' + \lambda\delta)^{m-1}.$$

Which gives

$$(p + \lambda)^m \mathcal{L}E(p) = 1.$$

But one has

$$Y(t)\frac{t^{m-1}}{(m-1)!}e^{-\lambda t} \ \sqsupset \ \frac{1}{(p+\lambda)^m}; \ \text{for } Re \ p > -Re \ \lambda.$$

Hence

$$E = T_{Y\frac{t^{m-1}}{(m-1)!}e^{-\lambda t}}.$$

We recapture the particular case of 3) of Example 4.4 in Chapter 4.

3) **Elementary solution of** $D^2 + \omega^2$, $\omega \in R$. We look for E in $\mathcal{D}'(\mathbb{R}_+)$ such that

$$(\delta^{(2)} + \omega^2\delta) * E = \delta.$$

Which gives

$$(p^2 + \omega^2)\mathcal{L}E(p) = 1.$$

But

$$\psi(t)\frac{\sin \omega t}{\omega} \ \sqsupset \ \frac{1}{p^2 + \omega^2} = \frac{1}{2i\omega}\left(\frac{1}{p - i\omega} + \frac{1}{p + i\omega}\right); \ \text{for } Re \ p > 0.$$

Hence

$$E = T_{Y\frac{\sin \omega t}{\omega}}.$$

4) **Elementary solution of a particular equation**

We look for E in $\mathcal{D}'(\mathbb{R}_+)$ such that

$$\left(\sum_{j=0}^{m}\alpha_j\delta^{(j)}\right) * E = \delta.$$

Which gives

$$\left(\sum_{j=0}^{m}\alpha_j p^j\right)\mathcal{L}E(p) = 1.$$

One has

$$\sum_{j=0}^{m}\alpha_j p^j = \alpha_m(p - z_1)\cdots(p - z_m)$$

where $z_1, ..., z_m$ are the roots of the polynomial $\alpha_m z^m + \cdots + \alpha_1 z + \alpha_0$. But for $0 \le k \le m$,

$$Ye^{z_k t} \;\sqsupset\; \frac{1}{p - z_k}; \quad Re\ p > -Re\ z_k.$$

Hence

$$E = T_{\frac{1}{\alpha_m}Ye^{z_1 t} * \cdots * Ye^{z_m t}}.$$

Remark 9.3. The method used pertains to the frame of symbolic calculus. For more details see [9] or [16].

The Laplace transformation has been used long ago by physicists as a fundamental tool in symbolic calculations, without rigorous justifications. It appears to be a generalization of the Fourier transformation to particular functions or distributions which are useful in the resolution of convolution equations. The images $\mathcal{L}(f)$ and $\mathcal{L}(T)$ of a function f or a distribution T are not always defined on the whole space $\mathbb{C}$ but only on a strip (Proposition 9.1 and Definition 9.2). Numerous examples are given to illustrate this fact.

Notice that the exchange and uniqueness theorems are the basic instruments in the process of symbolic calculus used in solving convolution equations. The method has been enlightened by known classical examples.

In [3], [14] and [15], the Laplace transformation is directly considered for distributions together with other notions. It is not mentioned in [6]. Due to its importance and keeping in mind that this is an introductory course, we isolated it. Moreover, we began with functions as in [10] to make its introduction more natural. This indicates from the beginning its very link with the Fourier transformation.

Our aim is to show the efficiency of the Laplace transformation. We presented only the bilateral one. For more readings on the transformation itself see [3], [10], [14] and [15]. Concerning the symbolic calculus and its applications see [9], [15] and [16].

Chapter 10

Introduction to the Theory of Kernels

In order to solve some equations, we are led to introduce different kinds of kernels (regular, regularizing and very regular). We associate to every differential operator with C^∞-coefficients a kernel said to be fundamental. We give the hypoellipticity theorem of Schwartz which allows to show the hypoellipticity of certain partial differential operators.

10.1 The notion of a kernel

Let $P(D)A = B$ be a Partial Differential Equation (PDE) with constant coefficients. If E is a fundamental solution, then the solution is given by $A = E * B$. If $B \in \mathcal{E}'(\Omega)$, one can consider the linear map $\mathcal{K}\colon T \longmapsto E * T$ of $\mathcal{E}'(\Omega)$ into $\mathcal{D}'(\Omega)$. So $A = \mathcal{K}(B)$. Moreover
$$P(D)(\mathcal{K}(T)) = T, \ \forall T \in \mathcal{E}'(\Omega).$$
Notice that if, without supposing the existence of a fundamental solution, there is a linear map $\mathcal{K}$ which satisfies the latter relation, then $A = \mathcal{K}(B)$ is the solution of $P(D)A = B$.

Now we examine the restriction of $\mathcal{K}$ to $\mathcal{D}(\Omega)$. If $\varphi \in \mathcal{D}(\Omega)$, then $E * \varphi \in \mathcal{D}'(\Omega)$ and for $\psi \in \mathcal{D}(\Omega)$
$$\langle \mathcal{K}(\varphi), \psi \rangle = \langle E \otimes \varphi, \psi \rangle.$$
Whence the following linear map:
$$\mathcal{D}(\Omega) \otimes \mathcal{D}(\Omega) \longrightarrow \mathbb{C}$$
$$\varphi \otimes \psi \longmapsto \langle E * \varphi, \psi \rangle.$$
It is continuous and hence extends as a distribution on $\Omega \times \Omega$, denoted by K. Thus we have
$$\langle K, \varphi \otimes \psi \rangle = \langle \mathcal{K}(\varphi), \psi \rangle ; \ \varphi, \psi \in \mathcal{D}(\Omega).$$

Conversely, a given $K \in \mathcal{D}'(\Omega \times \Omega)$ determines a linear map $\mathcal{K} : \mathcal{D}(\Omega) \longrightarrow \mathcal{D}'(\Omega)$ by

$$\langle \mathcal{K}(\varphi), \psi \rangle = \langle K, \varphi \otimes \psi \rangle.$$

We will see that this approach allows the resolution of PDE's with coefficients of class C^∞. The preceding leads us to the notion of a kernel.

In the sequel, U and V will be open subsets of $\mathbb{R}^p$ and $\mathbb{R}^q$ respectively.

Definition 10.1. We call a kernel K on $U \times V$ any distribution on $U \times V$. The kernel $\overleftrightarrow{K}$ on $U \times V$, given by

$$\left\langle \overleftrightarrow{K}, \theta \right\rangle = \left\langle K, \overleftrightarrow{\theta} \right\rangle,$$

where $\overleftrightarrow{\theta}(x, y) = \theta(y, x)$, is called the symmetric of K.

With any kernel K on $U \times V$ we associate canonically the linear map $\mathcal{K} : \mathcal{D}(V) \longrightarrow \mathcal{D}'(U)$ defined by

$$\langle \mathcal{K}(\psi), \varphi \rangle = \langle K, \varphi \otimes \psi \rangle; \ \psi \in \mathcal{D}(V), \ \varphi \in \mathcal{D}(U).$$

We have $\mathcal{K}(\psi) \in \mathcal{D}'(U)$, for every $\psi \in \mathcal{D}(V)$. Moreover, $\mathcal{K}$ is clearly continuous if $\mathcal{D}'(U)$ is endowed with its weak topology (called also vague topology). In fact, it remains continuous for a stronger topology, that is the one of uniform convergence on the bounded subsets of $\mathcal{D}(U)$. The latter is defined by the family $(p_B)_B$ of seminorms, where B runs over the collection of bounded subsets of $\mathcal{D}(U)$ with

$$p_B(T) = \sup_{\varphi \in B} |\langle T, \varphi \rangle|.$$

Indeed it is sufficient to show that, for every compact set $L \subset V$, the restriction $\mathcal{K} : \mathcal{D}_L(V) \longrightarrow \mathcal{D}'(U)$ of $\mathcal{K}$ to $\mathcal{D}_L(V)$ is continuous. Now for a bounded subset B of $\mathcal{D}(U)$, there is a compact set $H \subset U$ such that $B \subset \mathcal{D}_H(U)$. The continuity of K on $\mathcal{D}_{H \times L}(U \times V)$ shows that there is a constant c and an integer $m \in \mathbb{N}$ such that

$$|\langle K, \theta \rangle| \leq c \, p_{H \times L, m}(\theta); \ \theta \in \mathcal{D}_{H \times L}(U \times V).$$

From this we deduce that

$$p_B(\mathcal{K}(\psi)) = \sup_{\varphi \in B} |\langle K, \varphi \otimes \psi \rangle|; \ \psi \in \mathcal{D}(V), \varphi \in \mathcal{D}(U)$$

$$\leq c \sup_{\varphi \in B} p_{H,m}(\varphi) p_{L,m}(\psi).$$

Whence the assertion since $\sup\limits_{\varphi \in B} p_{H,m}(\varphi)$ is finite by hypothesis. In an analogous manner, one defines the linear map $\overset{\leftrightarrow}{\mathcal{K}}$ associated to the kernel $\overset{\leftrightarrow}{K}$ by

$$\left\langle \overset{\leftrightarrow}{\mathcal{K}}(\varphi), \psi \right\rangle = \left\langle \overset{\leftrightarrow}{K}, \psi \otimes \varphi \right\rangle.$$

We also remark that

$$\left\langle \overset{\leftrightarrow}{\mathcal{K}}(\varphi), \psi \right\rangle = \langle \mathcal{K}(\psi), \varphi \rangle; \ \varphi \in \mathcal{D}(U), \psi \in \mathcal{D}(V).$$

Example 10.1. If $f \in \mathcal{L}^1_{loc}(U \times V)$, then $K = T_f$ is a kernel on $U \times V$. For $\psi \in \mathcal{D}(V)$ and $\varphi \in \mathcal{D}(U)$, one has

$$\langle \mathcal{K}(\psi), \varphi \rangle = \langle T_f, \varphi \otimes \psi \rangle$$
$$= \int_{U \times V} f(x,y)\varphi(x)\psi(y)dxdy.$$

Using Fubini's theorem, one sees that $g(x) = \int_V f(x,y)\psi(y)dy$ exists, for almost every $x \in U$, and is in $\mathcal{L}^1_{loc}(U \times V)$. Then $\mathcal{K}(\psi) = T_g$. Notice that if $U = V$, then $\overset{\leftrightarrow}{K} = K$. Hence also $\overset{\leftrightarrow}{\mathcal{K}} = \mathcal{K}$. In particular if $U = V = \mathbb{R}$ and $f(x,y) = e^{-2i\pi xy}$, one has $g = \mathcal{F}(\psi)$.

Example 10.2. If $S \in \mathcal{D}'(U)$ and $T \in \mathcal{D}'(V)$, then $K = S \otimes T$ is a kernel on $U \times V$ and $\overset{\leftrightarrow}{K} = S \otimes T$. For $\psi \in \mathcal{D}(V)$, one has $\mathcal{K}(\psi) = \langle T, \psi \rangle S$ and for $\varphi \in \mathcal{D}(U)$, one has $\overset{\leftrightarrow}{K}(\varphi) = \langle S, \varphi \rangle T$.

We now introduce different kinds of kernels used in solving some *PDE.*'s.

10.2 Regular kernels

Let K be a kernel on $U \times V$, $\mathcal{K} : \mathcal{D}(V) \longrightarrow \mathcal{D}'(U)$ the linear map associated with K, and $\overset{\leftrightarrow}{\mathcal{K}} : \mathcal{D}(U) \longrightarrow \mathcal{D}'(V)$ the linear map associated with $\overset{\leftrightarrow}{K}$. Useful kernels are those for which the range of $\mathcal{K}$ is contained in

$$\mathcal{E}_1(U) = \{T_f, \ f \in \mathcal{E}(U)\}.$$

It is known that $\mathcal{K} : \mathcal{D}(V) \longrightarrow \mathcal{E}_1(U)$ is continuous when $\mathcal{E}_1(U)$ is endowed with the topology induced by the strong topology of $\mathcal{D}'(U)$. As the spaces $\mathcal{E}_1(U)$ and $\mathcal{E}(U)$ are algebraically isomorphic by $b : T_f \longmapsto f$, one considers on $\mathcal{E}_1(U)$ another topology coming from $\mathcal{E}(U)$. It is defined by the family $(p'_{K,m})_{K,m}$ of seminorms given by

$$p'_{K,m}(T_f) = p_{K,m}(f).$$

This topology is stronger than the topology of the uniform convergence on the bounded subsets of $\mathcal{D}(U)$. Indeed if B is a bounded subset of $\mathcal{D}(U)$ and K a compact set of U such that $B \subset \mathcal{D}_K(U)$ with B bounded in $\mathcal{D}_K(U)$ (Dieudonné-Schwartz lemma; cf. [15] T.1, p.60), then

$$p_B(T_f) \leq \left(\int_K |f(x)|\, dx \right) \sup_{\varphi \in B} p_{K,0}(\varphi), \forall f \in \mathcal{E}(U).$$

The map $\mathcal{K} : \mathcal{D}(V) \longrightarrow (\mathcal{E}_1(U), (p'_{K,m})_{K,m})$ is not continuous in general.

Definition 10.2. We say that $\mathcal{K}$ is

(1) semi-regular with respect to x, or on the left, if the map $\mathcal{K} : \mathcal{D}(V) \longrightarrow (\mathcal{E}_1(U), (p'_{K,m})_{K,m})$ is continuous.

(2) semi-regular with respect to y, or on the right, if the map $\overleftrightarrow{\mathcal{K}} : \mathcal{D}(U) \longrightarrow (\mathcal{E}_1(V), (p'_{K,m})_{K,m})$ is continuous.

(3) regular if it is semi-regular both on the left and on the right.

Let K be a semi-regular kernel on the left. The following map is linear and continuous

$$\widetilde{\mathcal{K}} = b \circ \mathcal{K} : \mathcal{D}(V) \longrightarrow \mathcal{E}_1(U) \longrightarrow \mathcal{E}(U).$$

Consider its transpose ${}^t\widetilde{\mathcal{K}} : \mathcal{E}'(U) \to \mathcal{D}'(V)$ which is linear and continuous. Then

$$\left\langle {}^t\widetilde{\mathcal{K}}(T), \psi \right\rangle = \left\langle T, \widetilde{\mathcal{K}}(\psi) \right\rangle;\ T \in \mathcal{E}'(U),\ \psi \in \mathcal{D}(V).$$

In particular, for every $\varphi \in \mathcal{D}(U)$, one has

$$\left\langle {}^t\widetilde{\mathcal{K}}(T_\varphi), \psi \right\rangle = \left\langle T_\varphi, \widetilde{\mathcal{K}}(\psi) \right\rangle$$

$$= \int_U \varphi(x)\widetilde{\mathcal{K}}(\psi)(x)dx$$

$$= \left\langle T_{\widetilde{\mathcal{K}}(\psi)}, \varphi \right\rangle$$

$$= \left\langle b^{-1}\left(b(\mathcal{K}(\psi))\right), \varphi \right\rangle$$

$$= \left\langle \mathcal{K}(\psi), \varphi \right\rangle.$$

But

$$\langle \mathcal{K}(\psi), \varphi \rangle = \left\langle \overleftrightarrow{\mathcal{K}}(\varphi), \psi \right\rangle$$

$$= \left\langle \left(\overleftrightarrow{\mathcal{K}} \circ b \right)(T_\varphi), \psi \right\rangle.$$

Hence

$$^t\widetilde{\mathcal{K}}(T_\varphi) = \left(\overleftrightarrow{\mathcal{K}} \circ b \right)(T_\varphi), \ \varphi \in \mathcal{D}(U).$$

So we have the following result.

Proposition 10.1. *Let K be a semi-regular kernel on the left on $U \times V$. Then the map $\overleftrightarrow{\mathcal{K}} \circ b$ extends to a continuous linear map of $\mathcal{E}'(U)$ into $\mathcal{D}'(V)$. It is nothing else than the transpose of the map $b \circ \mathcal{K}$.*

Example 10.3. Let $f \in \mathcal{L}^1_{loc}(U \times V)$ and consider the kernel $K = T_f$. We have seen that $\mathcal{K}(\psi) = T_g$ with $g \in \mathcal{L}^1_{loc}(U)$. Hence K is semi-regular on the left if and only if $g \in \mathcal{E}(U)$.

Example 10.4. For $S \in \mathcal{D}'(U)$ and $T \in \mathcal{D}'(V)$, consider the kernel $K = S \otimes T$ on $U \times V$. For $\psi \in \mathcal{D}(V)$, one has $\mathcal{K}(\psi) = \langle T, \psi \rangle S$. Hence K is semi-regular on the left if and only if $S \in \mathcal{E}_1(U)$. It is semi-regular on the right if and only if $T \in \mathcal{E}_1(V)$.

Example 10.5. If $T \in \mathcal{D}'(\mathbb{R}^n)$, one defines a kernel on $\mathbb{R} \times \mathbb{R}$ by

$$\langle K, \varphi \otimes \psi \rangle = \langle T * T_\psi, \varphi \rangle; \ \varphi \in \mathcal{D}(\mathbb{R}^n), \ \psi \in \mathcal{D}(\mathbb{R}^n).$$

Then

$$\mathcal{K}(\psi) = T * T_\psi \text{ and } \mathcal{K}(\varphi) = \overset{\vee}{T} * T_\varphi.$$

Let us show that K is regular on the left. It is clear that $\mathcal{K}$ is defined on $\mathcal{D}(\mathbb{R}^n)$ with values in $\mathcal{E}_1(\mathbb{R}^n)$, being also continuous. Indeed, one has $T * T_\psi = T_g$ with $g \in \mathcal{E}(\mathbb{R}^n)$ given by

$$g(t) = \langle T_x, \psi(t - x) \rangle = \left\langle T, \tau_t(\overset{\vee}{\psi}) \right\rangle.$$

We are going to show that, for every compact subset L of $\mathbb{R}^n$, $\psi \longmapsto T_g$ is a continuous map of $\mathcal{D}_L(\mathbb{R}^n)$ into $\mathcal{E}_1(\mathbb{R}^n)$, i.e., for every compact subset A of $\mathbb{R}$ and for every $m \in \mathbb{N}$, there are $c > 0$ and $l \in \mathbb{N}$ such that

$$p_{A,m}(g) \le c \, p_{L,l}(\psi).$$

Now if $t \in A$ and $supp\, \psi \subset L$, then $supp\, \tau_t(\overset{\vee}{\psi})$ is included in the compact set $H = A - L$. The continuity of T on $\mathcal{D}_H(\mathbb{R}^n)$ implies the existence of a constant $c > 0$ and a positive integer m_1 such that

$$|\langle T, \chi \rangle| \leq c\, p_{H,m_1}(\chi);\ \chi \in \mathcal{D}_H(\mathbb{R}^n).$$

In particular, if $\chi = D^\beta(\tau_t(\overset{\vee}{\psi}))$ then one has

$$\sup_{t \in A}|D^\beta g(t)| \leq c\mathrm{sup}p_{H,m_1}(D^\beta \tau_t(\overset{\vee}{\psi})),$$

whence

$$p_{A,m}(g) \leq c\, p_{A-H,m+m_1}(\psi).$$

10.3 Regularizing kernels

As in Proposition 10.1, if K is a semi-regular kernel on the right on $U \times V$ then the map $\mathcal{K} \circ b : \mathcal{D}_1(V) \longrightarrow \mathcal{D}(V) \longrightarrow \mathcal{E}_1(U)$ extends to a continuous linear map $\mathcal{E}'(V) \longrightarrow \mathcal{D}'(U)$. This extension is $^t(b \circ \overset{\leftrightarrow}{\mathcal{K}})$. It can have images in $\mathcal{E}_1(U)$.

We are led to the following notions:

Definition 10.3. Let K be a semi-regular kernel on the right (resp. on the left) on $U \times V$.

a) We say that K is regularizing on the right (resp. on the left) if the extension of $\mathcal{K} \circ b$ sends continuously $\mathcal{E}'(V)$ into $\mathcal{E}_1(U)$ (resp. the extension $^t(b \circ \mathcal{K})$ of $\overset{\leftrightarrow}{\mathcal{K}} \circ b$ sends continuously $\mathcal{E}'(U)$ into $\mathcal{E}_1(V)$).

b) We say that K is regularizing if it is both regularizing on the right and on the left.

Here is a general example of a regularizing kernel:

Proposition 10.2. *The kernel* $K = T_f$ *with* $f \in \mathcal{E}(U \times V)$ *is regularizing.*

Proof. We know that $\mathcal{K}(\psi) = T_g$ with

$$g(x) = \int_V f(x,y)\psi(y)dy.$$

By the theorem on derivation under the integral sign, one sees that $g \in \mathcal{E}(U)$. Hence K is semi-regular on the right. So it remains to show that the extension of $\mathcal{K} \circ b$ sends continuously $\mathcal{E}'(V)$ into $\mathcal{E}_1(U)$. For $T \in \mathcal{E}'(V)$ and $\varphi \in \mathcal{D}(U)$, one has

$$\left\langle {}^t(b \circ \overleftrightarrow{\mathcal{K}})(T), \varphi \right\rangle = \left\langle T, (b \circ \overleftrightarrow{\mathcal{K}})(\varphi) \right\rangle.$$

Now for $\psi \in \mathcal{D}(V)$,

$$
\begin{aligned}
\left\langle \overleftrightarrow{\mathcal{K}}(\varphi), \psi \right\rangle = \langle \mathcal{K}(\psi), \varphi \rangle &= \int_U g(x)\varphi(x)dx \\
&= \int_U \left(\int_V f(x,y)\psi(y)dy \right) \varphi(x)dx \\
&= \int_V \left(\int_U f(x,y)\varphi(x)dx \right) \psi(y)dy \\
&= \langle T_h, \psi \rangle
\end{aligned}
$$

with

$$h(y) = \int_U f(x,y)\varphi(x)dx.$$

Hence

$$
\begin{aligned}
\left\langle {}^t(b \circ \overleftrightarrow{\mathcal{K}})(T), \varphi \right\rangle &= \langle T, h \rangle \\
&= \langle T, \langle T_\varphi, f(.,.) \rangle \rangle \\
&= \langle T \otimes T_\varphi, f(.,.) \rangle \\
&= \langle T_\varphi, \langle T, f(.,.) \rangle \rangle \\
&= \langle T_F, \varphi \rangle
\end{aligned}
$$

with

$$F(x) = \langle T, f(x,.) \rangle.$$

Moreover, $F \in \mathcal{E}(U \times V)$, by Theorem 3.1 of Chapter 3. $\qquad \square$

10.4 Very regular kernels

Now we strengthen the notion of a regular kernel.

Definition 10.4. Let K be a kernel on $U \times U$. We say that K is very regular on U if

(i) K is regular.

(ii) For every $T \in \mathcal{E}'(U)$ and every open subset Ω of U, the restriction of $^t(b \circ \overleftrightarrow{K})(T)$ to Ω is in $\mathcal{E}_1(\Omega)$ whenever $T_{|\Omega} \in \mathcal{E}_1(\Omega)$.

Here is a general example of a very regular kernel.

Proposition 10.3. *Let K be a regular kernel on $U \times U$ and Δ^c the complement of the diagonal Δ in $U \times U$. Then K is very regular whenever $K_{|\Delta^c} \in \mathcal{E}_1(\Delta^c)$.*

Proof. Let $T \in \mathcal{E}'(U)$ and let Ω be an open subset of U such that $T_{|\Omega} \in \mathcal{E}_1(\Omega)$. We will show that the restriction of $S(T) = {}^t(b \circ \overleftrightarrow{K})(T)$ to Ω is in $\mathcal{E}_1(\Omega)$. It is sufficient to prove that, for any relatively compact subset W of Ω, the restriction of $S(T)$ to W is in $\mathcal{E}_1(W)$. Let W be such an open set and $\varphi_0 \in \mathcal{D}(U)$, equal to 1 on a neighborhood of $\overline{W}$ with $supp\ \varphi_0 \subset \Omega$. The distributions T, $\varphi_0 T$ and $(1 - \varphi_0)T$ have compact supports. Hence K being regular, $S(T)$, $S(\varphi_0 T)$ and $S[(1 - \varphi_0)T]$ have a meaning as distributions on U. Then, since S is linear, one gets

$$S(T) = S(\varphi_0 T) + S[(1 - \varphi_0)T].$$

Now $\varphi_0 T \in \mathcal{D}_1(\Omega) \subset \mathcal{D}_1(U)$ for $T_{|\Omega} \in \mathcal{E}_1(\Omega)$. Hence $S(\varphi_0 T) \in \mathcal{E}_1(U)$ since K is semi-regular on the left. On the other hand let $V = U \setminus \overline{W}$, the complement of $\overline{W}$ in U, and consider $H = K_{|W \times V}$. As $U \cap V = \emptyset$, $H \in \mathcal{E}_1(W \times V)$ by hypothesis. Hence it is regularizing on $W \times V$ by Proposition 10.2. Let $\mathcal{K}_1$ be the linear map, from $\mathcal{E}'(V)$ into $\mathcal{E}_1(W)$, which is associated to it. Put $T_1 = [(1 - \varphi_0)T]_{|V}$. It is an element of $\mathcal{E}'(V)$ and hence $\mathcal{K}_1(T_1) \in \mathcal{E}_1(U)$. But $\mathcal{K}_1(T_1)$ is exactly the restriction of $K[(1 - \varphi_0)T]$ to W. Hence $S[(1 - \varphi_0)T]_{|W}$ is an element of $\mathcal{E}_1(W)$. $\square$

10.5 Fundamental kernels of partial differential operators

Let Ω be an open subset of $\mathbb{R}$ and

$$P = P(D) = \sum_{|j| \leq m} a_j D^j \quad \text{where}\ \ j \in \mathbb{N}^n$$

a differential operator with C^∞-coefficients on Ω (cf. Section 2.7.3 of Chapter 2). Put

$$P(\varphi) = \sum_{|j| \leq m} a_j\ D^j \varphi,\ \forall \varphi \in \mathcal{E}(\varphi).$$

One obtains a continuous linear map of each of the spaces $\mathcal{D}(\Omega)$ and $\mathcal{E}(\Omega)$ into itself. For every $T \in \mathcal{D}'(\Omega)$, put

$$\mathbf{P}(T) = \sum_{|j| \leq m} a_j \, D^j T.$$

One obtains a continuous linear map of each of the spaces $\mathcal{D}'(\Omega)$ and $\mathcal{E}'(\Omega)$ into itself. Notice that for every $\varphi \in \mathcal{D}(\Omega)$, one has $\mathbf{P}(T_\varphi) = T P_{(\varphi)}$.

Definition 10.5.

1) Let K be a kernel on an open subset Ω of $\mathbb{R}$. It is said to be a fundamental kernel on the left (resp. on the right) for $\mathbf{P}$ if

$$(\mathcal{K} \circ b) \left[\mathbf{P}(T_\varphi) \right] = T_\varphi, \ \forall \varphi \in \mathcal{D}(\Omega),$$

respectively

$$\mathbf{P} \left[\mathcal{K} \circ b^{-1}(T_\varphi) \right] = T_\varphi, \ \forall \varphi \in \mathcal{D}(\Omega).$$

2) It is said to be fundamental if it is fundamental both on the left and on the right.

Remark 10.1.

1) If $\mathbf{P}$ admits a fundamental kernel on the left (resp. on the right), then it admits a left inverse and is one-to-one (resp. a right inverse and is onto).

2) A kernel K is fundamental on the left (resp. on the right) for $\mathbf{P}$ if

$$\mathcal{K}(P(\varphi)) = T_\varphi, \ \forall \varphi \in \mathcal{D}(\Omega),$$

respectively

$$\mathbf{P}(\mathcal{K}(\varphi)) = T_\varphi, \ \forall \varphi \in \mathcal{D}(\Omega).$$

Under additional hypotheses, the relations of Definition 10.5 extend to $\mathcal{E}'(\Omega)$.

Proposition 10.4. *Let K be a regular kernel on the right on Ω. If K is a fundamental kernel on the right (resp. on the left) for $\mathbf{P}$, then*

$$^t(b \circ \overleftrightarrow{\mathcal{K}})(\mathbf{P}\,(T)) = T, \ \forall T \in \mathcal{E}'(\Omega)$$

respectively

$$\mathbf{P}(\mathcal{K} \circ b^{-1}(T)) = T, \ \forall T \in \mathcal{E}'(\Omega).$$

Proof. This follows from the density of $\mathcal{D}(\Omega)$ in $\mathcal{E}'(\Omega)$ (cf. Proposition 4.8 of Chapter 4) and the continuity of $\mathcal{K}$. $\qquad\square$

Example 10.6 (Operators with constant coefficients). *Let E be a fundamental solution of a partial differential operator P with constant coefficients. According to Example 10.3 of Section 10.2, the kernel K defined by*

$$\langle K, \varphi \otimes \psi \rangle = \langle E * T_\psi, \varphi \rangle \, ; \; \varphi, \psi \in \mathcal{D}(\mathbb{R}^n)$$

is regular, being also fundamental for P. Indeed we have seen that $\mathcal{K}(\psi) = E * T_\psi$. Also we have for every $\varphi \in \mathcal{D}(\mathbb{R}^n)$,

$$\begin{aligned}
\mathcal{K}(P(\varphi)) &= E * T_{P(\varphi)} \\
&= E * (T_{P(\delta)*\varphi}) \\
&= E * (P(\delta) * T_\varphi) \\
&= T_\varphi.
\end{aligned}$$

So K is fundamental on the left. One sees, in the same way, that it is also fundamental on the right.

Example 10.7 (Operators with variable coefficients). *Let E be a fundamental solution of a partial differential operator P with constant coefficients. Then for any $a \in \Omega$, the distribution $E_a = E * \delta_a$ satisfies $\mathbf{P}(E_a) = \delta_a$. This leads to the following notion.*

Definition 10.6. Let $\mathbf{P}$ be a partial differential operator with C^∞-coefficients on Ω and $a \in \Omega$. We say that a distribution E_a is a fundamental solution of $\mathbf{P}$ at a if $\mathbf{P}(E_a) = \delta_a$.

If $\mathbf{P} = \sum_{|j| \leq m} a_j D^j$ is a partial differential operator with C^∞-coefficients on Ω, admitting at each point a of Ω a fundamental solution E_a such that $E_a = T_{f_a}$ with

$$g : (x, a) \longmapsto f_a(x) \text{ in } \mathcal{L}^1_{loc}(\Omega \times \Omega),$$

then the kernel $K = T_g$ is fundamental on the right for P. Indeed we have seen (Example 10.3) that $\mathcal{K}(\psi) = T_h$ with

$$h(x) = \int_\Omega g(x, a)\psi(a)da.$$

Putting

$$Q(T) = \sum_{|j| \leq m} (-1)^{|j|} D^j (a_j T)$$

one obtains for every $\varphi \in \mathcal{D}(\Omega)$,

$$\langle P(\mathcal{K}(\psi)), \varphi) \rangle = \langle \mathcal{K}(\psi), Q(\varphi) \rangle$$
$$= \int_\Omega \left(\int_\Omega g(x,a)\psi(a)da \right) Q(\varphi)(x) \ dx$$
$$= \int_\Omega \left(\int_\Omega g(x,a)Q(\varphi)(x)dx \right) \psi(a) \ da.$$

But

$$\int_\Omega g(x,a)Q(\varphi)(x)dx = \int_\Omega f_a(x)Q(\varphi)(x)dx$$
$$= \langle E_a, Q(\varphi) \rangle$$
$$= \langle PE_a, \varphi \rangle$$
$$= \varphi(a).$$

Consequently

$$\langle P(\mathcal{K}(\psi)), \varphi) \rangle = \int_\Omega \varphi(a)\psi(a)da$$
$$= \langle T_\psi, \varphi \rangle \, .$$

In the same way, one shows that it is fundamental on the left.

10.6 Hypoelliptic operators

Definition 10.7. We say that a partial differential operator $\mathbf{P}$ with C^∞-coefficients on Ω is hypoelliptic if, for every $T \in \mathcal{D}'(\Omega)$, and for every open subset U of Ω, $T \in \mathcal{E}_1(U)$, whenever $\mathbf{P}T \in \mathcal{E}_1(U)$.

It is clear that the set

$$\mathcal{N} = \{T \in \mathcal{D}'(\Omega) : \mathbf{P}T = 0\}$$

is a vector subspace of $\mathcal{D}'(\Omega)$ and that it is closed for the vague topology. If $\mathbf{P}$ is hypoelliptic, one has $\mathcal{N} \subset \mathcal{E}_1(\Omega)$.

Proposition 10.5. *If P is a hypoelliptic operator, then $\mathcal{N}$ is a Fréchet subspace of $\mathcal{E}_1(\Omega)$ for the topology induced by $\mathcal{E}(\Omega)$.*

Proof. Follows from the closedness of $\mathcal{N}$ in $\mathcal{E}_1(\Omega)$ for the vague topology and the fact that the latter is coarser than the one coming from $\mathcal{E}(\Omega)$. □

Theorem 10.1 (On Schwartz hypoellipticity). *Let Ω be an open subset of $\mathbb{R}^n$ and $\mathbf{P}$ a differential operator with C^∞-coefficients on Ω. If $\mathbf{P}$ admits a fundamental kernel on the left K which is very regular, then it is hypoelliptic on Ω.*

Proof. Let $T \in \mathcal{D}'(\Omega)$ and U an open subset of Ω such that $\mathbf{P}T \in \mathcal{E}_1(U)$. We have to prove that $T \in \mathcal{E}_1(U)$. Suppose first that $T \in \mathcal{E}'(U)$. Since K is fundamental on the left and regular on the right one has, according to Proposition 10.4, ${}^t(b \circ \overleftrightarrow{\mathcal{K}})(\mathbf{P}\,(T)) = T$. Conclude by the fact that K is very regular on Ω. Now let $T \in \mathcal{D}'(\Omega)$ and V a relatively compact open subset of U. Take $\varphi_0 \in \mathcal{D}(U)$, equal to 1 on a neighborhood of V. Then $\varphi_0\, T \in \mathcal{E}'(V) \subset \mathcal{E}'(\Omega)$ and $\varphi_0\, T = T$ on V. Hence it suffices to show that $\varphi_0\, T \in \mathcal{E}_1(V)$. But $\mathbf{P}(\varphi_0\, T)$ is the restriction to V of $\mathbf{P}T$. Hence $\mathbf{P}(\varphi_0\, T) \in \mathcal{E}_1(V)$. Then by the beginning of this proof, one has $\varphi_0\, T \in \mathcal{E}_1(V)$ and consequently $T \in \mathcal{E}_1(V)$. As V is arbitrary, $T \in \mathcal{E}_1(U)$. $\qquad\square$

In the case of constant coefficients, the statement is.

Corollary 10.1. *Let $\mathbf{P}$ be a partial differential operator on $\mathbb{R}^n$ with constant coefficients. If $\mathbf{P}$ admits a fundamental solution $E \in \mathcal{E}_1(\mathbb{R}^n \setminus \{0\})$, then it is hypoelliptic.*

Proof. We have seen (Example 10.6) that the kernel K defined by

$$\langle K, \varphi \otimes \psi \rangle = \langle E * T_\psi, \varphi \rangle \,; \ \varphi, \psi \in \mathcal{D}(\mathbb{R}^n)$$

is fundamental for $\mathbf{P}$. To show that it is very regular, we will use Proposition 10.3. One has

$$E * T_\varphi = T_g \ \text{with}\ g(t) = \langle E_x,\, \psi(t - x) \rangle .$$

Let $f \in \mathcal{E}(\mathbb{R}^n \setminus \{0\})$ such that $E = T_f$. Then

$$\begin{aligned}
\langle K, \varphi \otimes \psi \rangle &= \int g(t)\varphi(t)\,dt \\
&= \int\int f\,(t - y)\psi(y)\varphi(t)\,dy\,dt.
\end{aligned}$$

So $K = T_F$ with $F(t, y) = f(t - y)$. Now the restriction of F to the complement of the diagonal of $\mathbb{R}^n \times \mathbb{R}^n$ is of class C^∞ for $f \in \mathcal{E}(\mathbb{R}^n \setminus \{0\})$, by hypothesis. $\qquad\square$

In the case of a single variable, we have the following.

Corollary 10.2. *Any differential operator*

$$\mathbf{P} = D^m + a_{m-1}D^{m-1} + \cdots + a_1 D + a_0$$

with C^∞-coefficients on $\mathbb{R}$ is hypoelliptic.

Proof. Proceeding as in Proposition 4.15 of Chapter 4, one shows that $\mathbf{P}$ admits, at any point $a \in \mathbb{R}$, a fundamental solution E_a given by $E_a = T_{\tau_a} Y u_a$ where Y is the Heaviside function and u_a is the solution of

$$\begin{cases} P_u = 0 & \text{on } \mathbb{R} \\ u(a) = u'(a) = \cdots = u^{(m-2)}(a) = 0 \\ \qquad\qquad u^{(m-1)}(a) = 1. \end{cases}$$

Let K be the kernel defined by $K = T_g$ with $g(x,a) = Y(x-a)\, u(x)$. We know that it is fundamental on the left for P. To show that it is very regular, we will use Proposition 10.3. For this it suffices to prove that $(x,a) \longmapsto u_a(x)$ is of class C^∞ on $\mathbb{R}^2$. Consider m independent solutions $u_1, ..., u_m$ of the equation $\mathbf{P}u = 0$. It is known that

$$u_a(x) = C_1(a)u_1(x) + \cdots + C_m(a)u_m(x)$$

where

$$(C_1(a), \ldots, C_m(a)) = (W(a))^{-1}\, (0, 0, ..., 0, 1)$$

with

$$W(a) = \begin{pmatrix} u_1 & \cdots & u_m \\ u_1' & \cdots & u_m' \\ \vdots & \vdots & \vdots \\ u_1^{(m-1)} & \cdots & u_m^{(m-1)} \end{pmatrix}$$

the Wronskian matrix of $\mathbf{P}u = 0$. Moreover the maps $u_1 \cdots u_m$ are of class C^∞, that also holds true for $W(a)$ and $(W(a))^{-1}$. It is known, from the theory of Wronskian matrices that W is invertible everywhere since it is invertible at a. So the map $(x,a) \longmapsto u_a(x)$ belongs to $\mathcal{E}(\mathbb{R}) \otimes \mathcal{E}(\mathbb{R}) \subset \mathcal{E}(\mathbb{R} \times \mathbb{R})$. $\qquad\square$

Example 10.8.

1) **Cauchy-Riemann operators.** The following operators, said to be of Cauchy-Riemann, are hypoelliptic on $\mathbb{R}^2$.

$$\frac{\partial}{\partial \overline{z}} = \frac{1}{2}\left(\frac{\partial}{\partial x_1} + i\frac{\partial}{\partial x_2}\right)$$

and

$$\frac{\partial}{\partial z} = \frac{1}{2}\left(\frac{\partial}{\partial x_1} - i\frac{\partial}{\partial x_2}\right).$$

Let us for example show that $\frac{\partial}{\partial \bar{z}}$ is hypoelliptic. By Corollary 10.1, we will prove that it admits a fundamental solution $E \in \mathcal{E}_1(\mathbb{R}^2\backslash\{0\})$, i.e.,

$$\frac{\partial}{\partial \bar{z}}E = \delta$$

or

$$\frac{\partial}{\partial \bar{z}}\delta \times E = \delta.$$

We know that

$$\mathcal{F}\left(\frac{\partial}{\partial \bar{z}}\delta\right) = T_f \text{ with } f(x_1,x_2) = 2\pi i(x_1 + ix_2)$$

and that

$$\mathcal{F}(\delta) = T_g \text{ with } g \equiv 1.$$

Considering the Fourier transformation in the sense of ultradistributions (cf. 6.7, Chapter 6) and using 2) of Remark 6.20 in Chapter 6, one obtains

$$\mathcal{F}E = T_h \text{ with } h(x_1,x_2) = \frac{1}{2\pi i} \times \frac{1}{x_1 + ix_2}.$$

Since h is locally integrable and bounded for $\|x\| \geq 1$, the distribution T_h is tempered. Hence it is so for $\overline{\mathcal{F}}T_h$, so that we have

$$\frac{\partial}{\partial \bar{z}}\left(\overline{\mathcal{F}}T_h\right) = \delta.$$

Remark 10.2. One can show that $\overline{\mathcal{F}}T_h = T_{2ih}$.

2) **Laplace operator on $\mathbb{R}^2$.** The Laplace operator

$$\Delta = \frac{\partial^2}{\partial x_1^2} + \frac{\partial^2}{\partial x_2^2}$$

on $\mathbb{R}^2$ is hypoelliptic. Indeed

$$\Delta = 4\frac{\partial}{\partial z}\frac{\partial}{\partial \bar{z}}$$

Remark 10.3. One can show, using for example Green's formula, that $E = T_f$ with $f(x) = \dfrac{1}{2\pi} \log \|x\|$ is a fundamental solution of Δ on $\mathbb{R}^2$, while composition preserves also hypoellipticity.

3) The previous example generalizes to $n \geq 3$. Consider the operator

$$\Delta = \frac{\partial^2}{\partial x_1^2} + \frac{\partial^2}{\partial x_2^2} + \cdots + \frac{\partial^2}{\partial x_n^2}.$$

We will show that it admits a fundamental solution $E \in \mathcal{E}_1(\mathbb{R}^n \setminus \{0\})$, i.e.,

$$\Delta E = \delta$$

or

$$\Delta\delta * E = \delta.$$

It is known that

$$\mathcal{F}(\Delta\delta) = T_f \text{ with } f(x) = -4\pi^2 \|x\|^2$$

and that

$$\mathcal{F}(\delta) = T_g \text{ with } g \equiv 1.$$

Considering the Fourier transformation in the sense of ultradistributions and using 2) of Remark 6.20 in Chapter 6, one obtains

$$\mathcal{F}E = T_h, \text{ with } h(x) = \frac{1}{-4\pi^2 \|x\|^2}.$$

Since h is locally integrable and bounded for $\|x\| \geq 1$, the distribution T_h is tempered. It is then also so for $\overline{\mathcal{F}}T_h = E$, so that we have

$$\Delta\left(\overline{\mathcal{F}}T_h\right) = \delta.$$

Remark 10.4. One can show in various ways that

$$\overline{\mathcal{F}}T_h = T_F \text{ with } F(x) = \frac{-1}{(n-2)S_n \|x\|^{n-2}}$$

where S_n is the surface of the unit sphere in $\mathbb{R}$. It is known that $S_n = \dfrac{2\pi^{n/2}}{\Gamma(\frac{n}{2})}$ where Γ is the Eulerian function of the second kind, defined on $]0, +\infty[$ by

$$\Gamma(x) = \int_0^{+\infty} t^{x-1} e^{-t} dt.$$

4) **Hypoellipticity of the diffusion operator.** Consider the so-called diffusion operator

$$\mathbf{P} = \frac{\partial}{\partial t} - \Delta \text{ in } \mathbb{R}^{n+1}$$

where Δ is the Laplacian in $\mathbb{R}^n$, i.e.,

$$\mathbf{P} = \frac{\partial}{\partial t} - \left(\frac{\partial^2}{\partial x_1^2} + \cdots + \frac{\partial^2}{\partial x_n^2} \right), \quad (t, x) \in \mathbb{R}^n.$$

It is hypoelliptic. Indeed we will show that it admits a fundamental solution $E \in \mathcal{E}_1(\mathbb{R}^n \backslash \{0\})$, i.e.,

$$\mathbf{P}E = \delta$$

or

$$\mathbf{P}(\delta) * E = \delta.$$

It is known that

$$\mathcal{F}(\mathbf{P}(\delta)) = T_f \text{ with } f(t, x) = 2\pi i t + 4\pi^2 \|x\|^2$$

and that

$$\mathcal{F}\delta = T_g \text{ with } g \equiv 1.$$

Considering the Fourier transformation in the sense of ultradistributions and using **2)** of Remark 6.20 in Chapter 6, one obtains

$$\mathcal{F}E = T_h \text{ with } h(t, x) = \frac{1}{2\pi i t + 4\pi^2 \|x\|^2}.$$

Now, h being locally integrable and bounded for $\|(t, x)\| \geq 1$, the distribution T_h is tempered. Hence it is also so for $\overline{\mathcal{F}}T_h$, so that one has

$$\mathbf{P}\left(\overline{\mathcal{F}}T_h \right) = \delta.$$

Remark 10.5. One can show that $\overline{\mathcal{F}}T_h = T_F$ with

$$F(t, x) = \frac{Y(t)}{(4\pi t)^n} \exp\left(\frac{-\|x\|^2}{4t} \right)$$

where Y is the Heaviside function.

5) **Non hypoellipticity of the wave operator.** Consider the so-called
wave operator defined on $\mathbb{R}^2$ by

$$\mathbf{P} = \frac{\partial^2}{\partial x^2} - \frac{\partial^2}{\partial y^2}.$$

It is not hypoelliptic. Indeed we will show that there exists $T \in \mathcal{D}'(\mathbb{R}^2)$
such that $\mathbf{P}T = 0$ with $T \notin \mathcal{E}_1(\mathbb{R}^2)$. For $T \in \mathcal{D}'(\mathbb{R}^2)$ and every $\varphi \in$
$\mathcal{D}(\mathbb{R}^2)$, one has

$$\langle \mathbf{P}T, \varphi \rangle = \left\langle T, \frac{\partial^2 \varphi}{\partial x^2} - \frac{\partial^2 \varphi}{\partial y^2} \right\rangle.$$

It is known that the rotation of angle $\pi/4$, in the positive sense allows
to obtain another expansion of $\mathbf{P}$. Indeed, let $\theta : \mathbb{R}^2 \longrightarrow \mathbb{R}^2$ be the
automorphism defined by

$$\theta(u, v) = (x, y) = \left(\frac{1}{2}(u, v), \frac{1}{\sqrt{2}}(u - v) \right).$$

One has

$$\frac{\partial^2 \varphi}{\partial^2 x^2} - \frac{\partial^2 \varphi}{\partial^2 y^2} = 2\frac{\partial^2 \varphi \circ \theta}{\partial u \, \partial v}.$$

Then

$$\langle \mathbf{P}T, \varphi \rangle = \left\langle T, \frac{2\partial^2 \varphi \circ \theta}{\partial u \, \partial v} \right\rangle = \left\langle \frac{2\partial^2 \theta(T)}{\partial u \, \partial v}, \varphi \right\rangle$$

where $\theta(T)$ is the distribution on $\mathbb{R}^2$ defined by

$$\langle \theta(T), \varphi \rangle = \langle T, \varphi \circ \theta \rangle, \quad \varphi \in \mathcal{D}(\mathbb{R}^2).$$

So the equation $\mathbf{P}(T) = 0$ is equivalent to

$$\frac{\partial^2 (\theta(T))}{\partial u \, \partial v} = 0.$$

Since θ is an automorphism, the whole issue is reduced to show that
the operator

$$Q = \frac{\partial^2}{\partial x \, \partial y}$$

is non hypoelliptic. Hence we are led to consider the equations

$$\frac{\partial T}{\partial x} = 0 \text{ and } \frac{\partial T}{\partial y} = 0; \ T \in \mathcal{D}'(\mathbb{R}^2).$$

If $T \in \mathcal{D}'(\mathbb{R}^2)$ is such that $\frac{\partial T}{\partial x} = 0$, then

$$0 = \left\langle \frac{\partial T}{\partial x}, \varphi \otimes \psi \right\rangle = \left\langle T, \frac{\partial}{\partial x}(\varphi \otimes \psi) \right\rangle; \ \varphi, \psi \in \mathcal{D}(\mathbb{R}).$$

For $\psi \in \mathcal{D}(\mathbb{R})$, the distribution S given by
$$\langle S, \varphi \rangle = \langle T, \varphi \otimes \psi \rangle \,; \ \varphi \in \mathcal{D}(\mathbb{R}),$$
is such that $S' = 0$. According to Proposition 2.11 of Chapter 2, there exists a complex constant $c(\psi)$ such that $S = T_g$ with $g \equiv c(\psi)$, i.e.,
$$\langle S, \varphi \rangle = \langle T, \varphi \otimes \psi \rangle = c(\psi) \int_{\mathbb{R}} \varphi(t) dt.$$
Taking $\varphi_0 \in \mathcal{D}(\mathbb{R})$ with $\int_{\mathbb{R}} \varphi_0(t) dt = 1$, one has $c(\psi) = \langle T, \varphi_0 \otimes \psi \rangle$. The map $A : \psi \longmapsto c(\psi)$ is then a distribution and we have $T = T_1 \otimes A$. To finish, observe that $T_1 \otimes B$ is a solution of $QT = 0$ for every $B \in \mathcal{D}'(\mathbb{R})$ and that $T_1 \otimes \delta \notin \mathcal{E}_1(\mathbb{R}^2)$ since its support has a void interior.

The theory of kernels is the general frame for the resolution of partial differential equations. Only its first elements are presented. The notions of a kernel K and its canonical associated linear map $\mathcal{K}$ are drawn out (cf. Section 10.1), for $PDE's$ with constant coefficients, in terms of a fundamental solution. In general, the range of $\mathcal{K}$ is a subspace of distributions. Particular ranges are very useful in practice. This leads to the introduction of different notions of kernels (regular, regularizing and very regular).

Of particular interest is the equation
$$P(D)T = 0, \quad T \in \mathcal{D}'(\Omega),$$
where $P(D) = \sum_{|j| \leq m} a_j D^j$ is a differential operator with C^∞-coefficients on Ω. The notion of a fundamental kernel for P is essential. If this is moreover very regular then P is hypoelliptic on Ω (hypoellipticity theorem of Schawrtz). Then the space of solutions is contained in $\mathcal{E}_1(\Omega)$.

If P has constant coefficients on $\mathbb{R}^n$, hypoellipticity is ensured by the existence of a fundamental solution $E \in \mathcal{E}_1(\mathbb{R}^n \setminus \{0\})$. The Cauchy-Riemann, the Laplace and the diffusion operators fit in well here.

The results can be found in most of the books on distributions. The difference lies in the presentation. Here, the notions are approached via concrete and classical situations.

We have reintroduced the notation T_f for a regular distribution, to avoid a probable confusion. Hopefully this will help the reader to concentrate on the new notions and conduct calculations with confidence.

The chapter contains only the first aspects of the theory of kernels. The purpose is to show the usefulness of distribution theory in applications. For more readings see [6], [14] and [15].

Chapter 11

Sobolev Spaces

We consider, for every $m \in \mathbb{N}$, the classical Hilbert space $H^m(\Omega)$ of Sobolev and the closure $H_0^m(\Omega)$ of $\mathcal{D}_1(\Omega) = \{T_f : f \in \mathcal{D}(\Omega)\}$ in $H^m(\Omega)$. In order to introduce, for $s \in \mathbb{R}$, the Sobolev space $H^s(\mathbb{R}^n)$, one is led to study the Hilbert space $L^2(\mu_s)$ of measurable (classes of) functions f such that $x \longmapsto |f(x)|(1 + ||x||^2)^{\frac{s}{2}}$ is square summable. We also present the injection theorem of Sobolev. Next the compact injection theorem, which includes Rellich theorem, in particular.

11.1 The spaces $H^m(\Omega)$, $m \in \mathbb{N}$

In this chapter we will often identify a function with its (equivalence) class.

We know that if $g \in L^2(\mathbb{R}^n)$, then T_g is a tempered distribution (cf. 4) of Examples 6.1, Chapter 6). Hence $T_g = \overline{\mathcal{F}}T$ with $T \in \mathcal{S}'(\mathbb{R}^n)$. Then according to Proposition 8.5 of Chapter 8, one has $T = T_{\mathcal{F}g}$. Moreover $\mathcal{F}g \in L^2(\mathbb{R}^n)$, by the Riesz-Plancherel theorem (Proposition 8.2, Chapter 8). Hence we are led to consider the space
$$H^0(\mathbb{R}^n) = \left\{T \in \mathcal{S}'(\mathbb{R}^n) : T = T_f; \ f \in L^2(\mathbb{R}^n)\right\}.$$
Clearly it is a vector subspace of $\mathcal{S}'(\mathbb{R}^n)$, isomorphic to $L^2(\mathbb{R}^n)$. Observe that it is not stable by derivation (take for f the characteristic function $\chi_{[0,1]}$). So we come to the following definition.

Definition 11.1. For $m \in \mathbb{N}$, the Sobolev space of order m on Ω, denoted $H^m(\Omega)$, is given by
$$H^m(\Omega) = \left\{T \in \mathcal{D}'(\Omega) : D^j T = T_{f_j}, \ f_j \in L^2(\Omega); \ |j| \leq m\right\}.$$

Remark 11.1. For every $m \in \mathbb{N}^*$, the space $H^m(\Omega)$ is embedded in $L^2(\Omega)$. Moreover

$$\mathcal{D}(\Omega) \subset ... \subset H^{m+1}(\Omega) \subset H^m(\Omega) \subset ... \subset H^0(\Omega) \simeq L^2(\Omega).$$

The scalar product of $L^2(\Omega)$ allows to define a scalar product in $H^m(\Omega)$. Indeed for $S, T \in H^m(\Omega)$, one has, for $|j| \leq m$, $D^j T = T_{f_j}$, $f_j \in L^2(\Omega)$ and $D^j S = T_{g_j}$, $g_j \in L^2(\Omega)$. Now, we put

$$< D^j T, D^j S > = < f_j, g_j >$$

and

$$< T, S >_m = \sum_{|j| \leq m} < D^j T, D^j S >,$$

whence, the associated norm

$$\|T\|_m = \left(\sum_{|j| \leq m} < D^j T, D^j T > \right)^{\frac{1}{2}} = \left(\sum_{|j| \leq m} \|f_j\|_2 \right)^{\frac{1}{2}}.$$

Putting

$$\|D^j T\|_{j,2} = \left(< D^j T, D^j T > \right)^{\frac{1}{2}},$$

one obtains

$$\|T\|_m = \left(\sum_{|j| \leq m} \left(\|D^j T\|_{j,2} \right)^2 \right)^{\frac{1}{2}}.$$

Definition 11.2. The topology defined by the norm $\|.\|_m$ is called the canonical topology of $H^m(\Omega)$.

It is clear that the injections $H^{m+1}(\Omega) \longrightarrow H^m(\Omega)$ are continuous. Moreover, a sequence $(T_p)_p$ in $H^m(\Omega)$ converges to zero in $H^m(\Omega)$ if and only if for every $|j| \leq m$, $(f_{j,p})_p$ tends to zero in $L^2(\Omega)$ where $D^j T_p = T f_{j,p}$.

Proposition 11.1. *For every $m \in \mathbb{N}$, the space $(H^m(\Omega), \|.\|_m)$ is a Hilbert space.*

Proof.   Let $(T_p)_p$ be a Cauchy sequence in $H^m(\Omega)$ with

$$D^j T_p = T f_{j,p} \in L^2(\Omega), \text{ for } |j| \leq m.$$

Then $(f_{j,p})_p$ is Cauchy in $L^2(\Omega)$ which is complete. Denote its limit by f_j. Putting $f = f_0$, one has

$$T_{f_0,p} \xrightarrow[p]{} T_f \text{ in } \mathcal{D}'(\Omega).$$

But D^j is continuous map of $\mathcal{D}'(\Omega)$ into $\mathcal{D}'(\Omega)$, hence

$$D^j T_{f_0,p} \xrightarrow[p]{} D^j T_f \text{ in } \mathcal{D}'(\Omega).$$

Now $D^j T_{f_0,p} = T_{f_j,p}$ which tends to $T f_j$ in $\mathcal{D}'(\Omega)$, whence $D^j T_f = T f_j$. So $T_f \in H^m(\Omega)$, and since $f_{j,p} \xrightarrow[p]{} f_j$ in $L^2(\Omega)$, one has

$$T_p \xrightarrow[p]{} T_f \text{ in } H^m(\Omega).$$

$\square$

We are now interested in density results. We know that $\mathcal{D}(\Omega)$ is dense in $\mathcal{D}'(\Omega)$ (Proposition 4.8 of Chapter 4). But $\mathcal{D}(\Omega)$ is embedded in $\mathcal{E}'(\Omega) \cap H^m(\Omega)$. Hence $\mathcal{E}'(\Omega) \cap H^m(\Omega)$ is dense in $\mathcal{D}'(\Omega)$. Since it is contained in $H^m(\Omega)$, one wonders if it is dense in $H^m(\Omega)$. Put

$$H_c^m(\Omega) = \mathcal{E}'(\Omega) \cap H^m(\Omega).$$

Proposition 11.2. *For every $m \in \mathbb{N}$, the space $H_c^m(\mathbb{R}^n)$ is dense in $H^m(\mathbb{R}^n)$.*

Proof. The proof is based on truncating. Let $\varphi \in \mathcal{D}(\mathbb{R}^n)$ such that $0 \leq \varphi \leq 1$ and $\varphi = 1$ on $\overline{B}(0,1)$. For every $p \in \mathbb{N}^*$, put $\varphi_p(x) = \varphi(\frac{x}{p})$. Then $(\varphi_p)_p$ is a truncating sequence on $\mathbb{R}^n$, associated with the closed balls $\overline{B}(0,p)$. For $T \in H^m(\mathbb{R}^n)$ put

$$T_p = \varphi_p T.$$

The sequence $(T_p)_p$ is in $\mathcal{E}'(\mathbb{R}^n)$. Let us show that it is in $H^m(\mathbb{R}^n)$. For $|j| \leq m$, $D^j T_p$ is a linear combination of terms of the form $D^\alpha \varphi_p \, D^\beta T$ with $\alpha + \beta = j$. As $T \in H^m(\mathbb{R}^n)$, each of these terms is of the form

$$T_{f_{\alpha,\beta,p}} \text{ with } f_{\alpha,\beta,p} \in L^2(\mathbb{R}^n).$$

To conclude, we show that

$$T_p \xrightarrow[p]{} T \text{ in } H^m(\mathbb{R}^n).$$

Or, to say it differently, for any $|j| \leq m$

$$f_{j,p} \xrightarrow[p]{} f_j \text{ in } L^2(\mathbb{R}^n)$$

where

$$D^j T_p = T_{f_{j,p}} \text{ and } D^j T = T_{f_j}.$$

On the one hand, for $j = 0$

$$f_{0,p} = \varphi_p f_0 \xrightarrow[p]{} f_0 \text{ in } L^2(\mathbb{R}^n),$$

by the Lebesgue theorem. On the other hand one also has, for every $i = 1, ..., n$

$$\frac{\partial T_p}{\partial x_i} = \left(\frac{\partial \varphi_p}{\partial x_i}\right) T + \varphi_p \frac{\partial T}{\partial x_i}.$$

The previous argument shows that

$$\varphi_p f_{j,i} \xrightarrow{p} f_{j,i} \text{ in } L^2(\mathbb{R}^n) \text{ where } \frac{\partial T}{\partial x_i} = T_{f_{j,i}}.$$

Let us show that

$$\frac{\partial T_p}{\partial x_i} f_0 \xrightarrow{p} 0 \text{ in } L^2(\mathbb{R}^n).$$

Since,

$$\left\|\frac{\partial \varphi_p}{\partial x_i}\right\|_\infty \leq \left\|\frac{\partial \varphi}{\partial x_i}\right\|_\infty,$$

we have

$$\int_{\mathbb{R}^n} \left|\left(\frac{\partial \varphi_p}{\partial x_i} f_0\right)(x)\right| dx \leq \left\|\frac{\partial \varphi}{\partial x_i}\right\|_\infty^2 \int_{\|x\|>p} |f_0(x)|^2 dx,$$

whence our assertion, for

$$\int_{\|x\|>p} |f_0(x)|^2 dx \xrightarrow{p} 0.$$

It follows that

$$f_{j,i,p} \xrightarrow{p} f_{j,i} \text{ in } L^2(\mathbb{R}^n) \text{ where } T_{f_{j,i,p}} = \frac{\partial T_p}{\partial x_i}.$$

Finally using the Leibniz formula, one shows by induction that $f_{j,p} \xrightarrow{p} f_j$ in $L^2(\mathbb{R}^n)$, for any $|j| \leq m$. $\quad\square$

Proposition 11.3. *For every* $m \in \mathbb{N}$, *the space* $\mathcal{D}(\mathbb{R}^n)$ *is dense in* $H^m(\mathbb{R}^n)$ *up to an isomorphism.*

Proof. We will show that $\{T_\varphi : \varphi \in \mathcal{D}(\mathbb{R}^n)\}$ is dense in $H_c^m(\mathbb{R}^n)$. The proof is based on regularization. Let $(\theta_p)_p$ be a regularizing sequence on $\mathbb{R}$. For $T \in H_c^m(\mathbb{R}^n)$, put $T_p = T * \theta_p$. Then $D^j T_p = T_{f_j} * \theta_p$, where $T_{f_j} = D^j T$ with $f_j \in L^2(\mathbb{R}^n)$. It is clear that $f_j * \theta_p$ is in $\mathcal{D}(\mathbb{R}^n)$. It remains to show that

$$T_p \xrightarrow{p} T, \text{ in } H^m(\mathbb{R}^n).$$

or to say it differently, for every $|j| \leq m$

$$f_j * \theta_p \xrightarrow{p} f_j \text{ in } L^2(\mathbb{R}^n).$$

But this results from the regularization theorem (cf. Chapter 1). $\quad\square$

We now consider an arbitrary open subset Ω of $\mathbb{R}^n$. It is known that $\mathcal{D}(\Omega)$ is embedded in $\mathcal{D}(\mathbb{R}^n)$ as follows: If $\varphi \in \mathcal{D}(\Omega)$, then extend it by zero outside of Ω. Denote by $\tilde{\varphi}$ this extension. So $\mathcal{D}(\Omega)$ is embedded in $H^m(\mathbb{R}^n)$.

Proposition 11.4. *For every $m \in \mathbb{N}$ and every $u \in \mathcal{D}(\Omega), T_{\tilde{u}} \in H^m(\mathbb{R}^n)$ and*

$$D^j T_{\tilde{u}} = T_{\widetilde{D^j u}}; \ |j| \leq m.$$

Proof. Let $\varphi \in \mathcal{D}(\mathbb{R}^n)$. We must show that

$$\langle D^j T_{\tilde{u}}, \varphi \rangle = \langle T_{\widetilde{D^j u}}, \varphi \rangle,$$

or

$$(-1)^{|j|} \langle T_{\tilde{u}}, D^j \varphi \rangle = \langle T_{\widetilde{D^j u}}, \varphi \rangle$$

i.e.,

$$(-1)^{|j|} \int_\Omega u(x) D^j \varphi(x) dx = \int_\Omega \varphi(x) D^j u(x) dx.$$

This equality is obtained integrating by parts. $\qquad \square$

Remark 11.2.

(1) For every $u \in \mathcal{D}(\Omega)$, one has

$$\|T_u\|_{H^m(\Omega)} = \|T_{\tilde{u}}\|_{H^m(\mathbb{R}^n)}$$

since

$$D^j T_{\tilde{u}} = T_{\widetilde{D^j u}} \quad \text{and} \quad D^j T_u = T_{D^j u}; \ |j| \leq m.$$

(2) For every $u \in \mathcal{D}(\Omega)$ and every $\varphi \in \mathcal{D}(\mathbb{R}^n)$, one has

$$\left| \int_\Omega u(x) D^j \varphi(x) dx \right|^2 \leq \int_\Omega \left| D^j \varphi(x) \right|^2 dx \int_\Omega |u(x)|^2 dx$$

$$\leq c_\varphi \|u\|^2_{L^2(\Omega)}$$

$$\leq c_\varphi \|T_u\|^2_m.$$

Similarly for

$$\left| \int_\Omega \varphi(x) D^j u(x) dx \right| \leq c'_\varphi \|T_u\|_m.$$

Hence the linear forms on $\mathcal{D}_1(\Omega) = \{T_u : u \in D(\Omega)\}$, given by

$$T_u \longmapsto (-1)^{|j|} \int_\Omega u(x) D^j \varphi(x) dx$$

and

$$T_u \longmapsto \int_\Omega \varphi(x) D^j u(x) dx$$

are continuous. Since they coincide on $\mathcal{D}_1(\Omega)$, they also coincide on its closure $\overline{\mathcal{D}_1(\Omega)}$ in $H^m(\Omega)$. The space $\overline{\mathcal{D}_1(\Omega)}$ is often denoted by $H_0^m(\Omega)$.

For $m = 0$, $H^0(\Omega)$ is isomorphic to $L^2(\Omega)$. Hence $\mathcal{D}(\Omega)$ is dense in $H^0(\Omega)$ up to an isomorphism (cf. Proposition 1.11 of Chapter 1). For $m \geq 1$, $\mathcal{D}_1(\Omega)$ is not always dense in $H^m(\Omega)$ as the following example shows.

Example 11.1. Let $\Omega =]0, 1[$ and let $f = 1$ on Ω. Then $T_f \in H^m(\Omega)$. But $T_f \notin \overline{\mathcal{D}_1(\Omega)}$. Indeed, by (2) of Remark 11.2, it suffices to show that $T_{\tilde{f}} \notin H^m(\mathbb{R})$. Now $\tilde{f} = \chi_\Omega$ the characteristic function of Ω and $(T_{\chi_\Omega})' = \delta_1 - \delta_0$ is not regular.

11.2 The spaces $H_0^m(\Omega)$, $m \in \mathbb{N}$

We denote by $H_0^m(\Omega)$ the closure of $\mathcal{D}_1(\Omega)$ in $H^m(\Omega)$. Endowed with the norm induced by $\|.\|_m$, it becomes a Hilbert subspace of $H^m(\Omega)$. Let us examine its orthogonal supplement denoted by $(H_0^m(\Omega))^\perp$. Let $T \in H^m(\Omega)$ with $D^j T = T_{f_j}$ and $f_j \in L^2(\Omega)$. But T is orthogonal to $H_0^m(\Omega)$ if and only if T is orthogonal to $\mathcal{D}_1(\Omega)$, by the density of $\mathcal{D}_1(\Omega)$ in $H_0^m(\Omega)$. Now for every $\varphi \in \mathcal{D}(\Omega)$,

$$\langle T, T_\varphi \rangle_m = \sum_{|j| \leq m} \langle f_j, D^j \varphi \rangle_{L^2(\Omega)}$$

$$= \sum_{|j| \leq m} \int f_j(x) \overline{D^j \varphi(x)} dx$$

$$= \sum_{|j| \leq m} \langle T_{f_j}, D^j \overline{\varphi} \rangle$$

$$= \sum_{|j| \leq m} (-1)^{|j|} \langle D^j T_{f_j}, \overline{\varphi} \rangle$$

$$= \sum_{|j| \leq m} (-1)^{|j|} \langle D^{2j} T, \overline{\varphi} \rangle.$$

Hence $T \in (H_0^m(\Omega))^\perp$ if and only if

$$\sum_{|j| \leq m} (-1)^{|j|} D^{2j} T = 0.$$

So we have the following result.

Proposition 11.5. *For every $m \in \mathbb{N}$, the orthogonal supplement of $H_0^m(\Omega)$ is the set of solutions in $H^m(\Omega)$ of the partial differential equation*

$$P(D)T = 0 \quad \text{where} \quad P(D) = \sum_{|j| \leq m} (-1)^{|j|} D^{2j}.$$

Remark 11.3.

(1) One can produce another proof of the fact that $\overline{\mathcal{D}_1(\mathbb{R}^n)} = H^m(\mathbb{R}^n)$. Indeed

$$(H_0^m(\mathbb{R}^n))^\perp = \left\{ T \in H^m(\mathbb{R}^n) : \sum_{|j| \leq m} (-1)^{|j|} D^{2j} T = 0 \right\}.$$

Using the formulas

$$\mathcal{F}(D^{2j}T) = (2i\pi M)^{2j} \mathcal{F}T, \quad |j| \leq m,$$

and the fact that $\mathcal{F}$ is one-to-one on $\mathcal{S}'(\mathbb{R}^n)$, one has

$$(H_0^m(\mathbb{R}^n))^\perp = \left\{ T \in H^m(\mathbb{R}^n) : \sum_{|j| \leq m} (-1)^{|j|} (2i\pi M)^{2j} \mathcal{F}T = 0 \right\}$$

$$= \left\{ T \in H^m(\mathbb{R}^n) : \left(\sum_{|j| \leq m} M^{2j} \right) \mathcal{F}T = 0 \right\}.$$

Given that

$$\sum_{|j| \leq m} M^{2j}(x) = \sum_{|j| \leq m} x_1^{2j_1} ... x_n^{2j_n} \neq 0 \text{ for every } x \in \mathbb{R}^n,$$

one obtains

$$(H_0^m(\mathbb{R}^n))^\perp = \{ T \in H^m(\mathbb{R}^n) : \mathcal{F}T = 0 \} = \{0\},$$

or

$$H_0^m(\mathbb{R}^n) = H^m(\mathbb{R}^n).$$

Remark 11.4. In the remainder of this section, we will examine the topological dual $(H_0^m(\Omega))'$ of $H_0^m(\Omega)$.

Remark 11.5.

(1) The space $(H_0^m(\Omega))'$ is embedded in $\mathcal{D}'(\Omega)$, since the injection $\mathcal{D}(\Omega) \longrightarrow H_0^m(\Omega)$ is continuous and with dense range.
(2) A distribution $T \in \mathcal{D}'(\Omega)$ is in $(H_0^m(\Omega))'$ if and only if it is continuous on $(\mathcal{D}(\Omega), \|.\|_m)$ where

$$\|\varphi\|_m = \|T_\varphi\|_m \quad \text{for every } \varphi \in \mathcal{D}(\Omega).$$

As $H_0^m(\Omega)$ is a Hilbert space, it is isomorphic to its dual considered as a Hilbert space. We explicit this isomorphism, using the differential operator $P(D)$.

Proposition 11.6. *The operator*

$$P(D) = \sum_{|j| \le m} (-1)^{|j|} D^{2j}$$

is an isometric isomorphism of $H_0^m(\Omega)$ onto $(H_0^m(\Omega))'$.

Proof.   We first show that $P(D)$ sends $H_0^m(\Omega)$ into $(H_0^m(\Omega))'$. Let $S \in H^m(\Omega)$ with $D^j S = T_{f_j}, f_j \in L^2(\Omega), |j| \le m$ and consider

$$T = P(D)S = \sum_{|j| \le m} (-1)^{|j|} D^{2j} S \in \mathcal{D}'(\Omega).$$

We claim that T is continuous on $(\mathcal{D}(\Omega), \|.\|_m)$. Indeed for every $\varphi \in \mathcal{D}(\Omega)$, one has

$$
\begin{aligned}
\langle T, \varphi \rangle &= \sum_{|j| \le m} (-1)^{|j|} \langle D^{2j} S, \varphi \rangle \\
&= \sum_{|j| \le m} \langle D^j S, D^j \varphi \rangle \\
&= \sum_{|j| \le m} \langle T_{f_j}, D^j \varphi \rangle \\
&= \sum_{|j| \le m} \langle f_j, D^j \overline{\varphi} \rangle_{L^2(\Omega)} \\
&= \langle S, T_{\overline{\varphi}} \rangle_m,
\end{aligned}
$$

hence

$$|\langle T, \varphi \rangle| \le \|S\|_m \|\varphi\|_m,$$

i.e., the continuity of T. Actually $P(D)$ is an isometry. Indeed we have obtained above that

$$\langle P(D)S, \varphi \rangle = \langle S, T_{\overline{\varphi}} \rangle_m,$$

hence

$$\|P(D)S\|_{(H_0^m(\Omega))'} = \sup_{\|\varphi\|_m \le 1,\ \varphi \in \mathcal{D}(\Omega)} |\langle P(D)S, \varphi \rangle|$$

$$= \sup_{\|\varphi\|_m \le 1,\ \varphi \in \mathcal{D}(\Omega)} |\langle S, T_{\overline{\varphi}} \rangle_m|$$

$$= \|S\|_m.$$

It remains to show that $P(D)$ is onto. Let $L \in (H_0^m(\Omega))'$. By the isomorphism theorem of Fréchet-F. Riesz, there exits a unique $S \in H_0^m(\Omega)$ such that

$$L(T) = \langle T, S \rangle_m \text{ for every } T \in H_0^m(\Omega).$$

If $D^j S = T_{g_j}$, $g_j \in L^2(\Omega)$, $|j| \le m$, one has for every $\varphi \in \mathcal{D}(\Omega)$

$$L(T_\varphi) = \langle T_\varphi, S \rangle_m$$

$$= \sum_{|j| \le m} \langle D^j T_\varphi, D^j S \rangle_m$$

$$= \sum_{|j| \le m} \langle T_{D^j \varphi}, T_{g_j} \rangle_m$$

$$= \sum_{|j| \le m} \langle D^j \varphi, g_j \rangle_{L^2(\Omega)}$$

$$= \sum_{|j| \le m} \langle T_{\overline{g}_j}, D^j \varphi \rangle.$$

If we consider $\overline{S} \in H_0^m(\Omega)$ such that $D^j \overline{S} = T_{\overline{g}_j}$, $|j| \le m$, we obtain

$$L(T_\varphi) = \langle P(D)\overline{S}; \varphi \rangle, \quad \varphi \in \mathcal{D}(\Omega),$$

or

$$L_{|\mathcal{D}_1(\Omega)} = P(D)\left(\overline{S} \circ b\right)$$

where b is defined by $T_\varphi \longmapsto \varphi$. Now $\overline{S} \circ b$ is continuous on $(\mathcal{D}_1(\Omega), \|.\|_m)$ and one gets $L = P(D)\left(\widetilde{\overline{S} \circ b}\right)$ where $\widetilde{\overline{S} \circ b}$ is the extension of $\overline{S} \circ b$ to $H_0^m(\Omega)$. $\square$

Here are some consequences.

Corollary 11.1. *For every $m \in \mathbb{N}$, the space $\mathcal{D}(\Omega)$ is dense in $(H_0^m(\Omega))'$ up to an isomorphism.*

Proof. It results from the fact that the range of a dense subset by an isometry which is onto is also dense.

$\square$

Observe that if $T \in (H_0^m(\Omega))'$ then, by the isomorphism theorem, there is $S \in H_0^m(\Omega)$ such that $T = P(D)S$.

If $D^j S = T_{g_j}$, $g_j \in L^2(\Omega)$, $|j| \leq m$, then one has

$$T = \sum_{|j| \leq m} (-1)^{|j|} D^{2j} S$$

$$= \sum_{|j| \leq m} (-1)^{|j|} D^j T_{g_j}$$

$$= \sum_{|j| \leq m} D^j T_{(-1)^{|j|} g_j}$$

$$= \sum_{|j| \leq m} D^j T_{f_j} \text{ with } f_j \in L^2(\Omega).$$

In fact, we thus have a characterization of the elements of $(H_0^m(\Omega))'$.

Corollary 11.2 (Structure of the elements of $(H_0^m(\Omega))'$). *For a distribution $T \in \mathcal{D}'(\Omega)$ to be in $(H_0^m(\Omega))'$, it is necessary and sufficient that T be of the form*

$$T = \sum_{|j| \leq m} D^j T_{f_j}, \; f_j \in L^2(\Omega).$$

Proof. It remains to show that if

$$T = \sum_{|j| \leq m} D^j T_{f_j} , \; f_j \in L^2(\Omega),$$

then it is continuous on $(\mathcal{D}'(\Omega), \|.\|_m)$. But for $\varphi \in \mathcal{D}(\Omega)$, one has

$$\langle T, \varphi \rangle = \sum_{|j| \leq m} (-1)^{|j|} \langle T_{f_j}, D^j \varphi \rangle$$

$$= \sum_{|j| \leq m} (-1)^{|j|} \langle f_j, D^j \overline{\varphi} \rangle_{L^2(\Omega)}.$$

Hence

$$|\langle T, \varphi \rangle| \leq \sum_{|j| \leq m} \|f_j\|_2 \|D^j \overline{\varphi}\|_2$$

$$\leq \|\varphi\|_m \sum_{|j| \leq m} \|f_j\|_2.$$

$\square$

Corollary 11.3. *For every $m \in \mathbb{N}$, one has the following injections.*

1) $S(\mathbb{R}^n) \subset H^m(\mathbb{R}^n) \subset S'(\mathbb{R}^n)$.
2) $S(\mathbb{R}^n) \subset (H^m(\mathbb{R}^n))' \subset S'(\mathbb{R}^n)$.

Proof.

1) The first injection follows from Proposition 6.6 of Chapter 6 and the second from 4) of Examples 6.1 of the same chapter.
2) As in 1), using the first assertion of Proposition 6.6 and 4) of Examples 6.1, of Chapter 6, as well as Corollary 11.1. $\qquad\square$

11.3 The spaces $L^2(\mu_s)$, $s \in \mathbb{R}$

By the very definition

$$T \in H^m(\mathbb{R}^n) \Longleftrightarrow (D^j T) = T_{f_j} \, ; \, f_j \in L^2(\mathbb{R}^n) \, ; \, |j| \leq m.$$

By the Parseval-Plancherel theorem,

$$T \in H^m(\mathbb{R}^n) \Longleftrightarrow \mathcal{F}(D^j T) = T_{g_j} ; g_j \in L^2(\mathbb{R}^n) \, ; \, |j| \leq m.$$

Since, $\mathcal{F}(D^j T) = (2i\pi M)^j \mathcal{F}T$, $|j| \leq m$, one obtains, putting $\mathcal{F}T = T_g$, $g \in L^2(\mathbb{R}^n)$,

$$T \in H^m(\mathbb{R}^n) \Longleftrightarrow (2i\pi M)^j T_g = g_j, \, g_j \in L^2(\mathbb{R}^n) \, ; \, |j| \leq m$$
$$\Longleftrightarrow (2i\pi M)^j g \in L^2(\mathbb{R}^n) \, ; \, |j| \leq m.$$

By Lemma 6.1 of Chapter 6,

$$T \in H^m(\mathbb{R}^n) \Longleftrightarrow \left(1 + \|x\|^2\right)^{\frac{m}{2}} g \in L^2(\mathbb{R}^n).$$

Or, otherwise, $T \in H^m(\mathbb{R}^n)$ if and only if $\mathcal{F}T = T_g$ with $g \in L^2(\mu_m)$, where μ_m is the measure $\left(1 + \|x\|^2\right)^{\frac{m}{2}} dx$. The latter is the measure with density $\left(1 + \|x\|^2\right)^{\frac{m}{2}}$ with respect to Lebesgue's measure dx.

The above motivates the following.

Definition 11.3. For $s \in \mathbb{R}$, we denote by $L_s^2(\mathbb{R}^n)$ the set (of classes) of measurable functions f such that

$$\int_{\mathbb{R}^n} |f(x)|^2 \left(1 + \|x\|^2\right)^s dx < +\infty.$$

The associated norm is denoted by $\|.\|_s$. In fact, it is a Hilbert space and we even have more.

Proposition 11.7. *For every r, $s \in \mathbb{R}$, the spaces $L^2(\mu_{r+s})$ and $L^2(\mu_r)$ are isometrically isomorphic. In particular, any $L^2(\mu_s)$ is isometrically isomorphic to $L^2(\mathbb{R}^n)$.*

Proof. For any function f, one has

$$|f(x)|^2 \left(1 + \|x\|^2\right)^{r+s} = \left(|f(x)| \left(1 + \|x\|^2\right)^{\frac{s}{2}}\right)^2 \left(1 + \|x\|^2\right)^r, \ x \in \mathbb{R}^n.$$

Consider the map

$$L^2(\mu_{r+s}) \longrightarrow L^2(\mu_r)$$
$$\overline{f} \longmapsto \overline{f \left(1 + \|M\|^2\right)^{\frac{s}{2}}}$$

where $\overline{f}$ is the equivalence class of f. It is an isomorphism. Moreover

$$\left\|\overline{f}\right\|_{r+s} = \left\|\overline{f(1 + \|M\|^2)^{\frac{s}{2}}}\right\|_r,$$

by the previous equality. For $r = 0$, one has $L^2(\mu_s) \simeq L^2(\mu_0) = L^2(\mathbb{R}^n)$, whence the second assertion.

$\square$

As a consequence, one gets:

Corollary 11.4. *For every $s \in \mathbb{R}$, the injection $\mathcal{S}(\mathbb{R}^n) \longrightarrow L^2(\mu_s)$ is continuous with dense range.*

Proof. Let $f \in \mathcal{S}(\mathbb{R}^n)$. The verification of $\overline{f} \in L^2(\mu_s)$ amounts to

$$\overline{f \left(1 + \|M\|^2\right)^{\frac{s}{2}}} \in L^2(\mathbb{R}^n).$$

For this, it suffices to show that

$$f \left(1 + \|M\|^2\right)^{\frac{s}{2}} \in \mathcal{S}(\mathbb{R}^n)$$

i.e.,

$$\left(1 + \|M\|^2\right)^{\frac{s}{2}} \in \mathcal{O}_M(\mathbb{R}^n),$$

which is actually the case (cf. Proposition 6.9 of Chapter 6). This injection is continuous as the composite of three continuous maps. The first is the multiplication by $\left(1+\|M\|^2\right)^{\frac{s}{2}}$ of $\mathcal{S}(\mathbb{R}^n)$ into $\mathcal{S}(\mathbb{R}^n)$, the second the injection of $\mathcal{S}(\mathbb{R}^n)$ into $L^2(\mathbb{R}^n)$ and the third the isometry

$$\overline{f} \longmapsto f\left(1+\|M\|^2\right)^{\frac{-s}{2}} \text{ of } L^2(\mathbb{R}^n) \text{ onto } L^2(\mu_s).$$

The density of the range comes from the fact that the injection of $\mathcal{S}(\mathbb{R}^n)$ into $L^2(\mathbb{R}^n)$ has a dense range. This is due to $\mathcal{D}(\mathbb{R}^n) \subset \mathcal{S}(\mathbb{R}^n)$ and the density of $\mathcal{D}(\mathbb{R}^n)$ in $L^2(\mathbb{R}^n)$, by Proposition 1.11 of Chapter 1. $\qquad\square$

Remark 11.6. For every $s \in \mathbb{R}$, the space $L^2(\mu_s)$ is embedded in $\mathcal{S}'(\mathbb{R}^n)$. Indeed it suffices to show that the measure T_f, $f \in L^2(\mu_s)$, is slowly increasing [cf. 2), Examples 6.1, Chapter 6]. Whence the existence of $k \in \mathbb{N}^*$ such that

$$\int_{\mathbb{R}^n} \frac{|f(x)|}{(1+\|x\|^2)^k}dx < +\infty.$$

Since $f \in L^2(\mu_s)$, we can write

$$\int_{\mathbb{R}^n} \frac{|f(x)|(1+\|x\|^2)^{\frac{s}{2}}}{(1+\|x\|^2)^{k-\frac{s}{2}}}dx < +\infty.$$

Just take $k \geq \frac{n}{2} + \frac{s}{2}$ and use the Cauchy-Schwarz inequality.

We now examine the dual of $L^2(\mu_s)$. Let $L \in \left(L^2(\mu_s)\right)'$. By the Fréchet-F. Riesz theorem, there exists a unique $g \in L^2(\mu_s)$ such that

$$L(f) = \langle f, g\rangle_{L^2(\mu_s)} \text{ for every } f \in L^2(\mu_s).$$

Let $T = L_{|_{\mathcal{S}(\mathbb{R}^n)}}$. For every $\varphi \in \mathcal{S}(\mathbb{R}^n)$, one has

$$L(\varphi) = \langle T, \varphi\rangle$$
$$= \langle \varphi, g\rangle_{L^2(\mu_s)}$$
$$= \int_{R^n} \varphi(x)\overline{g}(x)\left(1+\|x\|^2\right)^s ds$$
$$= \left\langle T_{\overline{g}(1+\|M\|^2)^s}, \varphi\right\rangle,$$

whence $T = T_{\overline{g}(1+\|M\|^2)^s}$. But for every $\varphi \in \mathcal{S}(\mathbb{R}^n)$, one has

$$\left|\left\langle T_{\overline{g}(1+\|M\|^2)^s}, \varphi\right\rangle\right| \leq \|g\|_s\|\varphi\|_s.$$

Hence $T_{\overline{g}(1+\|M\|^2)^s}$ is continuous on $(\mathcal{S}(\mathbb{R}^n), \|.\|_s)$. Furthermore, we have

$$L = T_{\overline{g}(1+\|M\|^2)^s}.$$

Now $g \in L^2(\mu_s)$ is equivalent to

$$\bar{g}\left(1 + \|M\|^2\right)^s \in L^2(\mu_{-s}).$$

So

$$\left(L^2(\mu_s)\right)' \subset \left[L^2(\mu_{-s})\right]_1,$$

where

$$\left[L^2(\mu_{-s})\right]_1 = \left\{T_h : h \in L^2(\mu_{-s})\right\}.$$

Conversely let T_h with $h \in L^2(\mu_{-s})$. It is continuous on $(\mathcal{S}(\mathbb{R}^n), \|.\|_s)$ for

$$|\langle T_h, \varphi \rangle| \leq \|h\|_{-s} \|\varphi\|_s ; \ \varphi \in \mathcal{S}(\mathbb{R}^n).$$

Hence $T_h \in \left(L^2(\mu_s)\right)'$ by Corollary 11.4. Finally one has

$$\left(L^2(\mu_s)\right)' = \left[L^2(\mu_{-s})\right]_1.$$

Moreover the isomorphism $f \longmapsto T_f$ of $L^2(\mu_{-s})$ onto $\left(L^2(\mu_s)\right)'$ is an isometry. Indeed for every $f \in L^2(\mu_{-s})$, one has

$$\begin{aligned}
\|T_f\|_{(L^2(\mu_s))'} &= \sup_{\|\varphi\|_s \leq 1, \ \varphi \in \mathcal{S}(\mathbb{R}^n)} |\langle T_f, \varphi \rangle| \\
&= \sup_{\|\varphi\|_s \leq 1, \ \varphi \in \mathcal{S}(\mathbb{R}^n)} \left| \int_{\mathbb{R}^n} f(x)\varphi(x)dx \right| \\
&= \sup_{\|\varphi\|_s \leq 1, \ \varphi \in \mathcal{S}(\mathbb{R}^n)} \left| \left\langle f\left(1 + \|M\|^2\right)^{-\frac{s}{2}}, \varphi\left(1 + \|M\|^2\right)^{\frac{s}{2}} \right\rangle \right|_{L^2(\mathbb{R}^n)} \\
&= \left\| f\left(1 + \|M\|^2\right)^{-\frac{s}{2}} \right\|_2 = \|f\|_{-s}.
\end{aligned}$$

So we have obtained the following result:

Proposition 11.8. *For every $s \in \mathbb{R}$, the hilbertian dual of $L^2(\mu_s)$ is isometrically isomorphic to the space $L^2(\mu_{-s})$.*

11.4   The spaces $H^s(\mathbb{R}^n)$, $s \in \mathbb{R}$

As for $m \in \mathbb{N}$, put for every $s \in \mathbb{R}$

$$H^s(\mathbb{R}^n) = \left\{T \in \mathcal{S}'(\mathbb{R}^n) : \mathcal{F}T = T_f, \ f \in L^2(\mu_s)\right\}.$$

To say it otherwise,

$$H^s(\mathbb{R}^n) = \overline{\mathcal{F}}\left(\left\{T_f, \ f \in L^2(\mu_s)\right\}\right).$$

In the sequel, $H^s(\mathbb{R}^n)$ will be endowed with the following associated scalar product and norm. For T, S in $H^s(\mathbb{R}^n)$ with $\mathcal{F}T = T_f$ and $\mathcal{F}S = T_g$, where f and g are in $L^2(\mu_s)$, put

$$\langle T, S \rangle_s = \langle f, g \rangle_{L^2(\mu_s)} = \int_{\mathbb{R}^n} f(x)\overline{g}(x) \left(1 + \|x\|^2\right)^s dx$$

and

$$\|T\|_s^2 = \|f\|_{L^2(\mu_s)}^2 = \int_{\mathbb{R}^n} |f(x)|^2 \left(1 + \|x\|^2\right)^s dx.$$

Remark 11.7. The Fourier transformation is an isometric isomorphism of $H^s(\mathbb{R}^n)$ onto $\left[L^2(\mu_s)\right]_1$, where

$$\left[L^2(\mu_s)\right]_1 = \left\{T_f : f \in L^2(\mu_s)\right\}.$$

Here are some properties of the spaces $H^s(\mathbb{R}^n)$.

Using Proposition 11.7, one obtains the following:

Proposition 11.9. *The convolution by* $\mathcal{F}\left[\left(1 + \|M\|^2\right)^s\right]$ *sends isometrically the space* $H^{r+s}(\mathbb{R}^n)$ *onto the space* $H^r(\mathbb{R}^n)$, *for any* r *and* s *in* $\mathbb{R}$. *In particular, all the spaces* $H^s(\mathbb{R}^n)$ *are isometrically isomorphic Hilbert spaces.*

Proposition 11.10.

i) *For every* $s \in \mathbb{R}$ *the map* $f \longmapsto T_f$, *of* $\mathcal{S}(\mathbb{R}^n)$ *into* $H^s(\mathbb{R}^n)$, *is one-to-one, continuous and with dense range.*

ii) *For every* $s \in \mathbb{R}$, *the strong dual of* $H^s(\mathbb{R}^n)$ *is continuously embedded in the strong dual of* $\mathcal{S}(\mathbb{R}^n)$.

iii) *For every* $s \in \mathbb{R}$, *the Hilbertian dual of* $H^s(\mathbb{R}^n)$ *is isometrically isomorphic to* $H^{-s}(\mathbb{R}^n)$.

Proof.

i) It is clear that the map at issue is one-to-one. Moreover one has

$$\|T_f\|_s^2 = \|f\|_{L^2(\mu_s)}^2 = \int_{\mathbb{R}^n} |f(x)|^2 \left(1 + \|x\|^2\right)^s dx.$$

Corollary 11.4 allows to conclude that it is continuous. The density of the range comes from the density of $S(R^n)$ in $L^2(\mu_s)$ (cf. Corollary 11.4).

ii) Results from i) and general results on transposition maps.

iii) Results from Proposition 11.8.

$\square$

11.5   Injection theorem of Sobolev

Let $\mathcal{C}_0(\mathbb{R}^n)$ be the space of continuous functions on $\mathbb{R}^n$ which vanish at infinity. Put

$$[\mathcal{C}_0(\mathbb{R}^n)]_1 = \{T_f : f \in \mathcal{C}_0(\mathbb{R}^n)\}.$$

Proposition 11.11. *Let $s \in \mathbb{R}$ be such that $s > \frac{n}{2}$. Then the space $H^s(\mathbb{R}^n)$ is continuously embedded in the space $[\mathcal{C}_0(\mathbb{R}^n)]_1$.*

Proof.   We will show that $H^s(\mathbb{R}^n)$ is continuously embedded in $\mathcal{F}\left([L^1(\mathbb{R}^n)]_1\right)$ and conclude by Remark 8.4 of Chapter 8, since

$$\overline{\mathcal{F}}\left([L^1(\mathbb{R}^n)]_1\right) = [\overline{\mathcal{F}}\left(L^1(\mathbb{R}^n)\right)]_1.$$

Given that

$$H^s(\mathbb{R}^n) = \overline{\mathcal{F}}\left([L^2(\mu_s)]_1\right),$$

it is sufficient to show that $L^2(\mu_s)$ is continuously embedded in $L^1(\mathbb{R}^n)$. Let then $f \in L^2(\mu_s)$. Write

$$f(x) = \left(1 + \|x\|^2\right)^{\frac{s}{2}} f(x) \frac{1}{\left(1 + \|x\|^2\right)^{s/2}}.$$

Observe that the function

$$x \longmapsto \left(1 + \|x\|^2\right)^{\frac{s}{2}} f(x)$$

is in $L^2(\mathbb{R}^n)$, and that for $s > \frac{n}{2}$, the function

$$x \longmapsto \left(1 + \|x\|^2\right)^{-\frac{s}{2}}$$

is also in $L^2(\mathbb{R}^n)$. So f is the product of two elements of $L^2(\mathbb{R}^n)$, so that

$$\|f\|_1 \leq \left(\int_{\mathbb{R}^n} \frac{1}{\left(1 + \|x\|^2\right)^s} dx\right) \|f\|_s.$$

□

More generally, we have the following.

Theorem 11.1. *Let $k \in \mathbb{N}$ and $s \in \mathbb{R}$ such that $s > \frac{n}{2} + k$. Then the space $H^s(\mathbb{R}^n)$ is continuously embedded in $\left[\mathcal{C}_0^k(\mathbb{R}^n)\right]_1$, where*

$$\mathcal{C}_0^k(\mathbb{R}^n) = \left\{ f \in \mathcal{C}^k(\mathbb{R}^n) : D^\alpha f \in \mathcal{C}_0(\mathbb{R}^n), \, |\alpha| \le k \right\}$$

is endowed with the norm

$$\|f\|_k = \max_{|\alpha| \le k} \sup \left\{ |D^\alpha f(x)| : x \in \mathbb{R}^n \right\}.$$

Proof. Let $T \in H^s(\mathbb{R}^n)$ and $\alpha \in \mathbb{N}^n$ with $|\alpha| \le k$. Then

$$D^\alpha T \in H^{s-k}(\mathbb{R}^n).$$

As $s - k > \frac{n}{2}$, the previous proposition insures that

$$D^\alpha T \in [\mathcal{C}_0(\mathbb{R}^n)]_1.$$

Then, according to the Du Bois-Raymond lemma (cf. [15]),

$$T \in \left[\mathcal{C}_0^k(\mathbb{R}^n)\right]_1.$$

It remains to show that the injection is continuous. Let $T \in H^s(\mathbb{R}^n)$ and $g \in L^2(\mu_s)$ so that $T = \overline{\mathcal{F}}T_g$. We know that there is an $f \in \mathcal{C}_0^k(\mathbb{R}^n)$ such that $T = T_f$. On the other hand, for any $|\alpha| \le k$,

$$D^\alpha T \in \overline{\mathcal{F}}\left(\left[L^1(\mathbb{R}^n)\right]_1\right).$$

Hence, let $h_\alpha \in L^1(\mathbb{R}^n)$ such that $D^\alpha T = \overline{\mathcal{F}}T_{h_\alpha}$ and $h_0 = g$. We have for $|\alpha| \le k$,

$$|D^\alpha f(x)| \le \int_{\mathbb{R}^n} |h_\alpha(y)| \, dy.$$

But one has

$$T_{h_\alpha} = \mathcal{F}(D^\alpha T_f) = (2i\pi M)^\alpha T_g,$$

whence $h_\alpha = (2i\pi M)^\alpha g$. Now h_α can be written

$$h_\alpha = (2i\pi M)^\alpha \left(1 + \|M\|^2\right)^{-\frac{s}{2}} g \left(1 + \|M\|^2\right)^{\frac{s}{2}}.$$

It is the product of two elements of $L^2(\mathbb{R}^n)$, since $g \in L^2(\mu_s)$ and

$$\int_{\mathbb{R}^n} |(2i\pi x)^\alpha|^2 \left(1 + \|x\|^2\right)^{-s} dx \le 4\pi^2 \int_{\mathbb{R}^n} \frac{1}{\left(1 + \|x\|^2\right)^{-k+s}} dx.$$

The second integral is finite for $2(-k+s) > n$. Finally the Cauchy-Schwarz inequality gives

$$\int_{\mathbb{R}^n} |h_\alpha(y)| \, dy \le \left(4\pi^2 \int_{\mathbb{R}^n} \frac{1}{\left(1 + \|x\|^2\right)^{-k+s}} dx \right) \|T\|_s,$$

for any $|\alpha| \le k$. Whence the continuity. $\square$

Corollary 11.5. *We have the following relations.*

(i) $\bigcap\limits_{s\in\mathbb{R}} H^s(\mathbb{R}^n) \subset [\mathcal{E}(\mathbb{R}^n)]_1$.

(ii) $[\mathcal{D}(\mathbb{R}^n)]_1 = \bigcap\limits_{s\in\mathbb{R}} H^s_c(\mathbb{R}^n)$

where $H^s_c(\mathbb{R}^n) = \{T \in H^s(\mathbb{R}^n) : supp\ T\ is\ compact\}$.

11.6 The compact injection theorem

We will need the following general lemma:

Lemma 11.1. *Let* $s \in \mathbb{R}$. *Then, for any* x, y *in* $\mathbb{R}^n$
$$\left(1 + \|x+y\|^2\right)^s \le (1+\|x\|)^{2|s|}\left(1+\|y\|^2\right)^s.$$

Proof. It is sufficient to establish the inequality for $s = \pm 1$. If $s = 1$, one has
$$1 + \|x+y\|^2 = 1 + \|x\|^2 + 2xy + \|y\|^2$$
$$\le 1 + \|x\|^2 + 2\|x\|\|y\| + \|y\|^2$$
$$\le (1+\|x\|)^2(1+\|y\|)^2 \quad \text{for } \|y\| < 1 + \|y\|^2.$$
For $s = -1$, we come back to $s = 1$, putting $u = x + y$ and $v = -x$. $\square$

For $s \in \mathbb{R}$ and K a compact subset of $\mathbb{R}^n$, put
$$H^s_K(\mathbb{R}^n) = \{T \in H^s(\mathbb{R}^n) : supp\ T \subset K\}.$$
Notice that for $r \le s$, $H^s(\mathbb{R}^n) \subset H^r(\mathbb{R}^n)$.

Theorem 11.2. *The canonical injection of* $H^s_K(\mathbb{R}^n)$ *into* $H^r(\mathbb{R}^n)$ *is compact for any* $r < s$.

Proof. It is to be shown that from any sequence $(T_j)_j$ bounded by 1, in $H^s_K(\mathbb{R}^n)$, one can extract a Cauchy subsequence (for $H^r(\mathbb{R}^n)$ is complete). Let $(f_j)_j \subset L^2(\mu_s)$ such that $\mathcal{F}T_j = T_{f_j}$. We first show that there is a subsequence $(T_k)_k$ such that $(f_k)_k$ is uniformly convergent on any compact subset K of $\mathbb{R}^n$. Take $\chi \in \mathcal{D}(\mathbb{R}^n)$ equal to 1 on a neighborhood of K. As $T_j = \chi T_j$, one has $T_{f_j} = \mathcal{F}(\chi T_j)$. So the functions f_j are of class $\mathcal{C}^\infty$, by Proposition 6.23 of Chapter 6. Hence $T_{f_j} = \mathcal{F}\chi * T_{f_j}$ and (cf. Proposition 4.6, Chapter 4)
$$f_j(x) = \int_{\mathbb{R}^n} \mathcal{F}\chi(x-y)f_j(y)dy.$$

Multiplying by $\left(1+\|x\|^2\right)^{\frac{s}{2}}$, one obtains

$$\left(1+\|x\|^2\right)^{\frac{s}{2}} f_j(x) = \int_{\mathbb{R}^n} \left(1+\|x\|^2\right)^{\frac{s}{2}} \mathcal{F}\chi(x-y) f_j(y) dy,$$

whence, by the previous lemma,

$$\left(1+\|x\|^2\right)^{\frac{s}{2}} |f_j(x)| \leq \int_{\mathbb{R}^n} (1+\|x-y\|)^{|s|} |\mathcal{F}\chi(x-y)| \left(1+\|y\|^2\right)^{\frac{s}{2}} |f_j(y)| \, dy.$$

Now by the Cauchy-Schwarz inequality one has

$$\left(1+\|x\|^2\right)^s |f_j(x)|^2$$
$$\leq \int_{\mathbb{R}^n} (1+\|x-y\|)^{2|s|} |\mathcal{F}\chi(x-y)|^2 \, dy \int_{\mathbb{R}^n} \left(1+\|y\|^2\right)^s |f_j(y)|^2 \, dy$$
$$\leq \int_{\mathbb{R}^n} (1+\|u\|)^{2|s|} |\mathcal{F}\chi(u)| \, du \quad \text{for } \|f_j\|_s < 1.$$

Similarly

$$\left(1+\|x\|^2\right)^s |f_j'(x)|^2 \leq \int_{\mathbb{R}^n} (1+\|u\|)^{|s|} |(\mathcal{F}\chi)'(u)|^2 \, d\mu = c_1,$$

where c_1 is a constant independent of x and of j. Then by the mean value theorem, the sequence $(f_j)_j$ is equicontinuous. One obtains the subsequence $(f_k)_k$ by Ascoli's theorem and the fact that the topology of compact convergence on $\mathcal{C}(\mathbb{R}^n)$ is metrizable. Finally, let us show that the subsequence $(T_k)_k$ is Cauchy in $H^r(\mathbb{R}^n)$. Let $\varepsilon > 0$ and B a closed ball such that

$$\frac{\left(1+\|x\|^2\right)^{\frac{r}{2}}}{\left(1+\|x\|^2\right)^{\frac{s}{2}}} \leq \varepsilon \text{ for } x \notin B \text{ (since } r < s).$$

One has

$$\|T_k - T_l\|_r \leq \varepsilon \|T_k - T_l\|_s + \left(\int_B \left(1+\|x\|^2\right)^r |(f_k - f_l)(x)|^2 \, dx\right)^{\frac{1}{2}}.$$

The conclusion follows from $\|T_k - T_l\|_s \leq 2$ and the fact that $(f_k)_k$ converges uniformly on the compact set B. $\qquad\square$

Remark 11.8. Let Ω be a bounded open subset of $\mathbb{R}^n$ and $T \in \mathcal{D}'(\Omega)$. One extends T to $\mathbb{R}^n$, putting

$$\left\langle \widetilde{T}, \varphi \right\rangle = \langle T, \chi\varphi \rangle \, ; \, \varphi \in \mathcal{D}'(\mathbb{R}^n)$$

where $\chi \in \mathcal{D}(\mathbb{R}^n)$ is equal to 1 on the compact set $\overline{\Omega}$.

Corollary 11.6 (Rellich's Theorem). *Let Ω be a bounded open subset of $\mathbb{R}^n$. For every $m \in \mathbb{N}$, the injection of $H_0^m(\Omega)$ into $H_0^{m-1}(\Omega)$ is compact.*

Proof. Let $(T_j)_j$ be a sequence bounded by 1 in $H_0^m(\Omega)$ and let $(f_j)_j \subset L^2(\Omega)$ such that

$$D^\alpha T_{f_j} = T_{g_{\alpha,j}} \text{ with } g_{\alpha,j} \in L^2(\Omega).$$

In order to apply the previous theorem, we show that $(\widetilde{T}_j)_j \subset H^m(\mathbb{R}^n)$. But one has $\widetilde{T}_j = T_{\widetilde{f}_j}$ where $\widetilde{f}_j$ is the extension of f_j to $\mathbb{R}^n$ by zero outside of Ω. Now for $|\alpha| \leq m$,

$$\widetilde{D^\alpha T}_{f_j} = T_{\widetilde{g_{\alpha.j}}} \text{ where } \widetilde{g_{\alpha.j}} \in L^2(\mathbb{R}^n).$$

Moreover $\widetilde{D^\alpha T}_{f_j} = D^\alpha T_{\widetilde{f}_j}$, i.e., for every $\varphi \in \mathcal{D}(\mathbb{R}^n)$

$$(-1)^{|\alpha|} \int_\Omega f_j(x) D^\alpha \varphi(x) dx = \int_\Omega \varphi(x) g_{\alpha,j}(x) dx.$$

Indeed, according to Proposition 11.4, the linear forms

$$T_\psi \longmapsto \int_\Omega \varphi(x) D^\alpha \psi(x) dx$$

and

$$T_\psi \longmapsto (-1)^{|\alpha|} \int_\Omega \psi(x) D^\alpha \varphi(x) dx$$

are equal on $\mathcal{D}_1(\Omega)$. They are also continuous on $(\mathcal{D}_1(\Omega), \|.\|_m)$, since

$$\left| \int_\Omega \varphi(x) D^\alpha \psi(x) dx \right|^2 \leq \left(\int_\Omega |\varphi(x)|^2 dx \right) \left(\int_\Omega |D^\alpha \psi(x)|^2 dx \right)$$
$$\leq c \|D^\alpha \psi\|_2^2 \leq c \|T_\psi\|_m^2$$

and

$$\left| \int_\Omega \psi(x) D^\alpha \varphi(x) dx \right| \leq c' \|T_\psi\|_m^2.$$

As $K = \overline{\Omega}$ is compact and $(\widetilde{T}_j)_j \subset H_K^m(\mathbb{R}^n)$, the previous theorem allows the extraction of a subsequence $(\widetilde{T}_k)_k$ which converges in $H^{m-1}(\mathbb{R}^n)$. It is then convergent in $H_0^{m-1}(\Omega)$. $\qquad\square$

Sobolev spaces are very important in the resolution of differential partial equations. The situation becomes comfortable when one has at his disposal a Hilbert space. For the equation

$$\sum_{|j| \leq m} (-1)^{|j|} D^{2j}(T) = 0,$$

where $T = T_g$, $g \in L^2(\Omega)$, the space

$$H^m(\Omega) = \left\{ T \in \mathcal{D}'(\Omega) : D^j T = T_{f_j}, \, f_j \in L^2(\Omega), |j| \le m \right\}$$

appears naturally. Then the set of solutions in $H^m(\Omega)$ is exactly the orthogonal supplement of the closure $H_0^m(\Omega)$ of the space $\mathcal{D}_1(\Omega) = \{T_f : f \in \mathcal{D}(\Omega)\}$ in $H^m(\Omega)$, cf. Proposition 11.5. The space $H^m(\Omega)$ is called the Sobolev space of order m on Ω; it is a Hilbert space. Different properties of $H^m(\Omega)$ are given in relations with other spaces already encountered. Observe that the operator

$$P(D) = \sum_{|j| \le m} (-1)^{|j|} D^{2j}(T)$$

is an isometric isomorphism of $H_0^m(\Omega)$ onto its dual $(H_0^m(\Omega))'$. At the beginning of Section 3, a characterization of the space $H^m(\Omega)$ is given. It allows a natural introduction of the more general Sobolev spaces $H^s(\mathbb{R}^n)$, with $s \in \mathbb{R}$.

When working in $H^s(\mathbb{R}^n)$ the injection theorem of Sobolev indicates the regularity of the solutions in relation with the uniqueness. The compact injection theorem (Rellich theorem, in particular) is often used to prove the existence when the problem is reduced to convergence phenomena.

The theory of Sobolev spaces is so large, and the content of this chapter is just a very first initiation to the matter, as in [6]. More along with applications can be found in [3], [6], [7] and [15].

Bibliography

[1] N. Bourbaki, Espaces vectoriels topologiques, Chap. I-II, Actua. Sci. industr., 1189, 2éme edition, Hermann, Paris 1966.

[2] H. Cartan, Sur les classes de Haar, C. R. Acad. Sci. Paris 211(1940), 789-762. Oeuvres Vol.III, pp. 1020-1022.

[3] Y. Choquet-Bruhat, Distributions. Théorie et problèmes, Masson et Cie, Paris, 1973.

[4] I. M. Gelfand et G. E. Shilov, Generalized functions, Vol. 1-3, Academic Press, 1966-1968.

[5] I. M. Gelfand et G. E. Shilov, Les distributions, Ed. technico-litteraires, Moscou. Traduction française, Dunod, 1962.

[6] H. Hogbé-Nlénd, Distributions et bornologies, Notas Inst. Mat. Estat. Univ. Sao Paulo, Série Matematica, No 3, 1973.

[7] L. Hörmander, The Analysis of Linear Partial Differential Operators, T. 1-2-3-4, Springer-Verlag, 1983-1985.

[8] J. Horváth, Topological Vector Spaces and Distributions, Addison Wesley, 1966.

[9] J. Mikusinsky, Operational Calculus, Pergamon, 1959.

[10] F. Roddier, Distributions et transformation de Fourier, Septième tirage, McGraw-Hill, Paris, 1978.

[11] W. Rudin, Real and Complex Analysis, McGraw-Hill Book Company, New york, 1966.

[12] W. Rudin, Functional Analysis, McGraw-Hill Book Company, New york, 1973.

[13] L. Schwartz, Méthodes mathématiques pour les sciences physiques, Hermann, Paris, 1965.

[14] L. Schwartz, Théorie des distributions, ($2^{éme}$ éd), Hermann, Paris, 1966.

[15] K. Vo-Khac, Distributions, analyse de Fourier, opérateurs aux dérivées partielles (2 tomes), Vuibert, 1970.

[16] K. Yosida, Functional Analysis, Sixth Edition, Springer-Verlag 1980.

List of Symbols

Index